Applied Mathematical Sciences
Volume 94

Applied Mathematical Sciences

(continued following index)

Frank C. Hoppensteadt

Analysis and Simulation of Chaotic Systems

With 74 illustrations

Springer-Verlag
New York   Berlin   Heidelberg   London   Paris
Tokyo   Hong Kong   Barcelona   Budapest

Frank C. Hoppensteadt
College of Natural Sciences
Michigan State University
East Lansing, MI 48824-1115 USA

Editors

F. John
Courant Institute of
 Mathematical Sciences
New York University
New York, NY 10012
USA

J.E. Marsden
Department of
 Mathematics
University of California
Berkeley, CA 94720
USA

L. Sirovich
Division of
 Applied Mathematics
Brown University
Providence, RI 02912
USA

Library of Congress Cataloging-in-Publication Data
Hoppensteadt, F.C.
   Analysis and simulation of chaotic systems / Frank C.
Hoppensteadt.
      p.   cm. — (Applied mathematical sciences; 94)
   Includes bibliographical references and index.
   ISBN 0-387-97916-6
   1. Chaotic behavior in systems. I. Title. II. Series: Applied
mathematical sciences (Springer-Verlag New York Inc.); v. 94.
QA1.A647   vol. 94
[Q175.5.C45]
510 s—dc20
[003'.7]   92-33590

Printed on acid-free paper.

Production managed by Karen Phillips; manufacturing supervised by Vincent Scelta.
Typeset by Asco Trade Typesetting Ltd., Hong Kong.
Printed and bound by R.R. Donnelley and Sons, Harrisonburg, VA.
Printed in the United States of America.

9 8 7 6 5 4 3 2 1

ISBN 0-387-97916-6 Springer-Verlag New York Berlin Heidelberg
ISBN 3-540-97916-6 Springer Verlag Berlin Heidelberg New York

Acknowledgments

I thank all of the teachers and students whom I have encountered, and I thank my parents, wife Leslie, children, stepchildren and pets for the many insights to life and mathematics that they have given me—often unsolicited. I have published parts of this book in the context of research or expository papers done with co-authors, and I thank them for the opportunity to have worked with them. The work presented here was mostly derived by others, although parts of it I was fortunate enough to uncover for the first time. I hope the organization and presentation are useful and clear. I have been very pleased with the production of this work carried out by Springer-Verlag, and I thank them for their support in the preparation and production of this book. My work has been supported by various agencies and institutions including the University of Wisconsin, New York University and the Courant Institute of Mathematical Sciences, the University of Utah, Michigan State University, the National Science Foundation, ARO, ONR, and the AFOSR. This investment in me has been greatly appreciated, and the work in this book describes some outcomes of that investment. I thank these institutions for their support.

This book is dedicated to Joseph B. Keller and to John R. Kinney,
both of whom have maintained a good sense of humor over the 25 years
I have known them.

Contents

Introduction

Scientists often use ordinary language models to describe observed phenomena. These are precise where data are known and appropriately imprecise otherwise. Ordinary language modelers carve away unknown chunks as more data are collected. On the other hand, mathematical modelers formulate minimal models that produce results similar to what is observed, and they account for new information by modifying the minimal model. Ordinary language models describe phenomena with perhaps less prejudice about the underlying process, but mathematical models make problems accessible to the consistency checking and predictive powers of mathematical analysis and computer simulations.

Often mathematical models are quite complicated, but simple approximations can extract from them important aspects of the process being studied. A stellar example of this is the Michaelis-Menten approximation to enzyme-mediated chemical reactions by which a complex reaction is ultimately described by a single differential equation that identifies two useful parameters (saturation constant and uptake velocity) and so suggests what experimental data might be useful. Another example is Semenov's theory of explosions, by which a single differential equation can be extracted from among over twenty chemical rate equations modeling chain branched reactions to describe thresholds of pressure and temperature that will result in an explosion.

Mathematical analysis includes geometric methods, such as phase planes or vectors, and analytical methods, such as iterations, perturbations, and transforms. Geometric methods are especially important in helping us create mental images of dynamical processes, and they are particularly useful in low-dimensional cases. Analytical methods enable us to calculate precisely how solutions depend on data in the model. Some geometrical methods are presented here as background material, but the emphasis in this book is on analytical methods, especially perturbation and iteration.

We occupy regions in space and time between very small and very large and very slow and very fast. These intermediate space and time scales are perceptible to us, and mathematical analysis has helped us to perceive scales that are beyond our senses. For example, it is very difficult to "understand" electric and magnetic fields, and our intuition is based best on solutions to Maxwell's equations. Fluid flows are quite complicated, but our intuition is shaped by knowledge of solutions of the Navier-Stokes equations. Most realistic mathematical models of physical or biological phenomena are complicated, and there are mathematical methods that extract simplifications to highlight and elucidate certain aspects of the underlying process. Simpler representations extracted using mathematical analysis can be used to suggest new experiments and to corroborate the model.

The mathematical methods presented and used here grew from several diverse schools of thought. The Göttingen school of Felix Klein, David Hilbert, and Richard Courant, and carried forward in this country by Fritz John, James Stoker, and Kurt Friedrichs, developed many important ideas that reached beyond ordinary differential equation models and studied important problems in continuum mechanics and wave propagation. The Soviet groups led by A. Liapunov, B. Krylov, and A. Kolmogorov developed novel approaches to problems of stability theory, statistical physics, and celestial mechanics. The work of Poincare on dynamical systems continues to provide new directions for us, and the U.S. mathematicians derived from George Birkhoff and Norbert Wiener have contributed to these topics as well. Perturbation and iteration methods were important to all of their work.

Computer simulations have become quite convenient only recently. At present, it is possible to simulate complicated mathematical models using a small lap-top computer, and the simulation power available in parallel computers and transputers is only now reaching wide use. Even with this prodigious power, computer simulations are limited. For example, certain chemical reactions involving time scales that differ by a hundred orders of magnitude, simulation of fairly simple chemicals in the design of drugs, and stress analysis of large buildings all require computational capacity beyond what is available today.

The computer simulations presented here describe several critical computer experiments that produced important and interesting applications. Computer simulations have become important since the work in Los Alamos in the 1940s when various methods, including Monte Carlo methods, were developed and used. They were implemented by rooms full of comptoimeter operators whose work was coordinated by various levels of schedulers. These simulations are still important today.

Mathematical analysis and computer simulations of dynamical systems are necessary to understand if one must deal with mathematical models of physical and biological phenomena. Understanding them is important in constructing mathematical models and in revealing the secrets they embody.

The chapters of this book present a variety of mathematical methods for solving problems that are sorted by gross features (linear, free, and forced problems and regular and singular perturbations). However, interwoven throughout the book are topics that reappear in many, often surprising, incarnations:

1. *Perturbations*. Even the words used here cause some problems. For example, *perturb* means to throw into confusion, but its purpose here is to relate to a simpler situation. Perturbations usually involve the identification of parameters, which to my astonishment is often understood by students as perimeters, from their studies of geometry. Parameter identification might involve difficult mathematical preprocessing in applications. However, once this is done, the basic perturbation methods for dynamical systems are Taylor's method for approximating a smooth function by a polynomial and Laplace's method for the asymptotic evaluation of integrals. These lead to the implicit function theorem and variants of it, and to matching, averaging, and central-limit theorems. Adaptations of these methods to various other problems are described here. Two particularly useful methods are the method of averaging and the quasistatic-state approximation method. These are dealt with in detail in Chapters 7 and 8.

2. *Iterations*. Iterations are mathematical procedures that begin with a state vector and change it according to some rule. The same rule is applied to the new state, and so on, and a sequence of iterates _. the rule results. Fra Fibonacci in 1202 introduced a famous iteration that describes the dynamics of an age-structured population. In Fibonacci's case, a population was studied and geometric growth was deduced. Several iterations are studied here. First, Newton's method, which continues to be the paradigm for iteration methods, is studied. Next, we study Duffing's iterative method and compare the results with similar ones derived using perturbation methods. Finally, we study chaotic behavior that often occurs when quite simple functions are iterated. There has been a controversy of sorts (iteration versus perturbation); each has its advocates, and each approach is useful.

3. *Chaos*. The term was introduced in its present connotation by James Yorke and T.Y. Li in 1976. It is not a precisely defined concept, but it occurs in various physical and religious ideas. For example, Boltzmann used it in a sense that eventually resulted in ergodic theories for dynamical systems, and Poincare had a clear image of the chaotic behavior of dynamical systems that occurs when stable and unstable manifolds cross. The book of Genesis begins with Kaos, and philosophical issues about randomness also become involved in discussions of chaos. For the most part, chaos is used here to indicate behavior of solutions to dynamical systems that is highly irregular and usually unexpected. We study several problems that are known to exhibit chaotic behavior and present methods for uncovering and describing this behavior. Related to chaotic systems are the following:

a. The study of almost periodic functions and of generalized Fourier analysis.
b. Poincare's stroboscopic method, which is based on snapshots of a solution at fixed time intervals—"Chaos, illumined by flashes of lightning" [from Oscar Wilde in another context].

c. Fractals, which are chaotic curves. They have been studied since Weierstrass, Hausdorff, and Peano at the start of this century.

d. Catastrophes, which were introduced by Rene Thom [I.1] in the 1960s. They were shown to lead to chaotic dynamics as well [I.2–I.4].

These and many other useful and interesting aspects of chaos are described here.

4.   *Oscillations.* Oscillators play fundamental roles in our lives— "discontented pendulums that we are" [R.W. Emerson]. For example, most of the cells in our bodies live an oscillatory life in an oscillatory chemical environment. The study of pendulums gives great insight to oscillators, and we focus a significant effort here in studying them and similar physical and electronic devices.

One of the most interesting aspects of oscillators is their tendency to synchronize with other nearby oscillators. This had been observed by musicians dating back at least to the time of Aristotle, and eventually it was addressed as a mathematical problem by van Huygens in the 17th century. This phenomenon is referred to as phase locking, and it now serves as a fundamental ingredient in the design of communications and computer-timing circuits. Phase locking is studied here for a variety of different oscillator populations using the rotation vector method. For example, using the model nerve cells described here (called VCON), we model neural networks as being flows on high-dimensional tori. Phase locking occurs when the flow reduces to a knot on the torus for the original and all nearby systems.

5. *Stability.* The stability of physical systems is often described using energy methods. These methods have been adapted to more general dynamical systems by Liapunov and others. Although we do study linear and Liapunov stability properties of systems here, the most important stability concept used here is that of stability under persistent disturbances. This idea explains why mathematical results obtained for minimal models can often describe behavior of nearby noisy systems that are operating in noisy environments. For example, think of a metal bowl having a lowest point in it. A marble placed in the bowl will eventually move to the minimum point. If the bowl is now dented with many small craters or if small holes are put in it, the marble will still move to near where the minimum of the original bowl had been, and the degree of closeness can be determined from the size of the dents and holes. The dents and the holes introduce irregular disturbances to the system, but the dynamics of the marble are similar in both the simple (ideal) bowl and the imperfect (realized) bowl. Ideas of stability are intimately connected with perturbations.

The two major topics studied here are mathematical analysis and computer simulation of dynamical systems. Each has its uses, its strengths, and its deficiencies. Our mathematical analysis builds mostly on perturbation and iteration methods. They are usually difficult methods to use, but once they are understood, they can provide information about systems that is not otherwise available. Understanding them for the examples presented here lays a basis for one to use computer packages such as Maple, Mathematica, or Matlab to construct perturbation expansions. Analytical methods can explain regular behavior of noisy systems, they can simplify complicated systems with fidelity to real behavior, and they can go beyond the edges of practical computability in dealing with rapid chemical reactions or trace-element calculations.

Computer simulation replaces much of the work formerly done by mathematicians, and use can be made of sophisticated software packages. Simulations illustrate solutions of a mathematical model by describing a sample trajectory, or sample path, of the process. Sample paths can be processed in a variety of ways—plotting, calculating ensemble statistics, and so on. Simulations do not describe the dependence of solutions on model parameters, nor are their stability, accuracy, or reliability always assured. They do not deal well with chaotics or catastrophes—irregular or unexpected rapid changes in solutions—and it is usually difficult to determine when chaos lurks nearby.

Mathematical analysis makes possible computer simulations; conversely, computer simulations can help with mathematical analysis. New computer-based methods are being derived with parallelization of computations, simplification of models through preprocessing, and so on, and the future holds great promise for combined work of mathematical and computer-based analysis. There have been many successes to date, including the discovery and analysis of solitons.

The material in this book is not presented in order of increasing difficulty. The first three chapters provide background information for the last five chapters, where oscillation and perturbation techniques and examples are developed. We begin with three examples that are useful throughout the rest of the book. These are circuits and pendulums. Next, we describe linear systems and spectral decomposition methods for solving them. These involve finding eigenvalues of matrices and deducing how they are involved in the solution of a problem. In the second chapter, we study free oscillations beginning with descriptions of how periodic or almost periodic solutions can be found in nonlinear dynamical systems using methods ranging from Poincare and Bendixson's method for two differential equations to entropy methods for nonlinear iterations. The third chapter presents stability methods for studying nonlinear systems. Particularly important for later work is the method of stability under persistent disturbances.

The remainder of the book deals with methods of approximation and simulation. First, some useful algebraic and topological methods are described, followed by a study of the implicit function theorem and modifications and generalizations of it. Then, regular perturbation problems are studied, in which a small parameter is identified and the solutions can be constructed directly using the parameter. This is illustrated by several important problems in nonlinear oscillations, including Duffing's equation and nonlinear resonance.

In Chapter Seven, the method of averaging is presented. This is one of the most interesting techniques in all of mathematics. It is closely related to Fourier analysis, to central-limit theorems of probability theory, and to the dynamics of physical and biological systems in oscillatory environments. We describe here multitime methods, Bogoliuboff's transformation, and the method of phase-amplitude coordinates.

Finally, the method of quasistatic-state approximations is presented. This method has been around in various useful forms since 1900, and it has been called by a variety of names—the method of matched asymptotic expansions being among the most civil. It has been derived in some quite complicated ways and in some quite simple ways. The approach taken here is of quasi-static manifolds, which has a clear geometric flavor that can aid intuition.

In rough terms, averaging applies when a system involves rapid oscillations that are slowly modulated, and the quasistatic-state approximation method applies when solutions decay rapidly to a manifold on which motions are slower. When problems arise where both kinds of behavior occur, they can usually be unraveled. But, there are many important problems where neither of these methods apply, including diffraction problems in electromagnetic theory, stagnation points in fluid flows, flows in domains with sharp corners, and problems with intermittent rapid time scales.

I have taught courses based on this material in a variety of ways depending on the time available and the background of the students. Usually, the material is taught as a full year course for graduate students in mathematics and engineering, and then I cover the whole book. Other times I have taken more advanced students who have had a good course in ordinary differential equations directly to Chapters 5, 6, 7, and 8. A one quarter course is possible using, for example, the first and seventh or the first and eighth chapters. For the most part Chapters 1 to 3 are intended as background material for the later chapters, although they contain important computer simulations that I like to cover in all of my presentations of this material. A course in computer simulations could deal with sections from Chapters 2, 4, 7, and 8. The exercises also contain several simulations that have been interesting and useful.

The exercises are graded roughly in increasing difficulty in each chapter. Some are quite straightforward illustrations of material in the text, and others are quite lengthy projects requiring extensive mathematical analysis or computer simulation. I have tried to warn readers about more difficult problems with an asterisk where appropriate.

Students must have some degree of familiarity with methods of ordinary differential equations, for example, from a course based on Coddington and Levinson [I.5], Hale [I.6], or Hirsch and Smale [I.7]. They should also be competent with matrix methods and be able to use a reference text such as

Gantmacher [I.8]. Some familiarity with *Fear and Loathing in Las Vegas* [I.9] and *Interpretation of Dreams* [I.10] have also been found to be useful by some students.

Frank C. Hoppensteadt
East Lansing, Michigan
March 1992

Part I
An Introduction to Nonlinear Systems

1
Oscillations of Linear Systems

A linear system of ordinary differential equations has the form

$$dx/dt = A(t)x + f(t)$$

Given an N-dimensional vector f and an $N \times N$-dimensional matrix $A(t)$ of functions of t, we seek a solution vector $x(t)$. We write $x, f \in E^N$ and $A \in E^{N \times N}$ and sometimes $x' = dx/dt$ or $\dot{x} = dx/dt$.

Most of the methods used to study nonlinear problems grew out of methods for linear problems, so mastery of linear problems is essential for understanding nonlinear ones. Section 1.1 presents several examples of physical systems that are analyzed in this book. In Sections 1.2 and 1.3, we study linear systems where A is a matrix of constants. In Sections 1.4 and 1.5, we study systems where A is a periodic or almost periodic matrix, and in Section 1.6, we consider general linear systems.

1.1 Examples

The following examples illustrate typical problems in oscillations and perturbations, and they are referred to throughout this book. The first two examples describe electrical circuits and the third a mechanical system.

1.1.1 *Voltage-Controlled Oscillators*

Modern integrated circuit technology has had a surprising impact on mathematical models. Rather than the models becoming more complicated as the number of transistors on a chip increases, the mathematics in many cases has become dramatically simpler, usually by design. *Voltage-controlled oscillators* (VCOs) illustrate this nicely. A VCO is an electronic device that puts out a voltage in a fixed wave form, say V, but with a variable phase x that is controlled by an input voltage, V_{in}. The device is described by the circuit diagram in Figure 1.1. The voltages in this and other figures are measured relative to a common ground that is not shown. The output wave form V

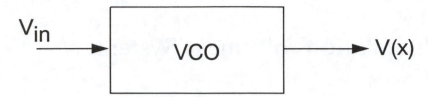

FIGURE 1.1. A voltage controlled oscillator. The controlling voltage V_{in} is applied to the circuit, and the output has a fixed wave form (V) whose phase x is modulated by V_{in}.

might be a square wave, a triangular wave, or a sinusoid, but its phase x is the unknown. VCOs are made up of many transistors, and a detailed model of the circuit is quite complicated [1.1, 1.2]. However, there is a simple input-output relation for this device: the input voltage, V_{in}, directly modulates the output phase as described by the equation

$$dx/dt = \omega + V_{in}$$

where the constant ω is called the *center frequency*. The center frequency, ω, is sustained by a separate (fixed supply) voltage in the device, and it can be changed by tuning resistances in the VCO. Thus, a simple differential equation models this device. The solution for x is found by integrating this equation:

$$x(t) = x(0) + \omega t + \int_0^t V_{in}(s)\, ds$$

$V(x)$ is observable in this circuit, and the higher the input voltage or the center frequency is, the faster V will oscillate.

Equations like this one for x play a central role in the theory of nonlinear oscillations. In fact, a primary goal is often to transform a given system into phase and amplitude coordinates, which is usually difficult to carry out. This model is given in terms of phase and serves as an example of how systems are studied once they are in phase and amplitude variables.

1.1.2 *Filters*

Filters are electrical circuits comprised of resistors, inductors, and capacitors. Figure 1.2 shows an *RLC* circuit, in which V_{in}, R, L, and C are given, and the unknowns are the output voltage (V) and the current (I) through the circuit. The circuit is described by

$$C\, dV/dt = I$$

$$L\, dI/dt + RI = V_{in} - V$$

The first equation describes the accumulation of charge in the capacitor; the

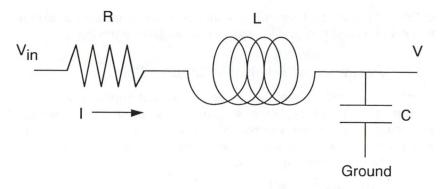

FIGURE 1.2. An RLC circuit.

second relates the voltage across the inductor and the voltage across the
resistor (Ohm's Law) to the total voltage $V_{in} - V$. Using the first equation to
eliminate I from the model results in a single second-order equation:

$$LC\frac{d^2V}{dt^2} + RC\frac{dV}{dt} + V = V_{in}$$

Thus, the first-order system for V and I can be rewritten as a second-order
scalar equation for V.

The *RLC* circuit is an example of a filter. In general, filters are circuits
whose models have the form

$$a_n\frac{d^nV}{dt^n} + a_{n-1}\frac{d^{n-1}V}{dt^{n-1}} + \cdots + a_0V = b_m\frac{d^mW}{dt^m} + \cdots + b_0W$$

where W is the input voltage, V is the output voltage, and the constants $\{a_i\}$
and $\{b_i\}$ characterize various circuit elements. Once W is given, this equation
must be solved for V.

Filters can be described in a concise form: Using the notation $p = d/dt$,
sometimes referred to as *Heaviside's operator*, we can write the filter equation
as

$$V = H(p)W$$

where the function H is a rational function of p:

$$H = \frac{(b_mp^m + \cdots + b_0)}{(a_np^n + \cdots + a_0)}$$

This notation is made precise later using Laplace transforms, but for now it
is taken to be a shorthand notation for the input-output relation of the filter.
The function H is called the filter's *transfer function*.

In summary, filters are circuits whose models are linear *n*th-order ordinary differential equations. They can be written concisely using the transfer function notation, and they provide many examples later in this book.

1.1.3   *Pendulum with Variable Support Point*

Simple pendulums are described by equations that appear in a surprising number of different applications. Consider a pendulum of length L supporting a mass m that is suspended from a point with vertical coordinate $V(t)$ and horizontal coordinate $H(t)$ as shown in Figure 1.3. The action integral for this mechanical system is defined by

$$\int_a^b \left[\frac{mL^2}{2} \left(\frac{dx}{dt} \right)^2 - \left(\frac{d^2V}{dt^2} + g \right) mL(1 - \cos x) - \frac{d^2H}{dt^2} mL \sin x \right] dt$$

where g is the acceleration of gravity. *Hamilton's principle* [1.3] shows that an extremum of this integral is attained by the solution $x(t)$ of the equation

$$L \frac{d^2x}{dt^2} + \left(\frac{d^2V}{dt^2} + g \right) \sin x + \frac{d^2H}{dt^2} \cos x = 0$$

which is the *Euler-Lagrange equation* for functions $x(t)$ that make the action integral stationary.

Furthrmore, a pendulum in a resistive medium to which a torque is ap-

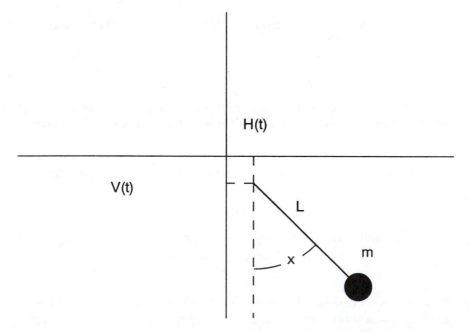

FIGURE 1.3. A pendulum with a moving support point. $(H(t), V(t))$ gives the location of the support point at time t. The pendulum is a massless rod of length L suspending a mass m. x measures the angular deflection of the pendulum from rest (down).

plied at the support point is described by

$$L\frac{d^2x}{dt^2} + f\frac{dx}{dt} + \left(\frac{d^2V}{dt^2} + g\right)\sin x + \frac{d^2H}{dt^2}\cos x = I$$

where f is the coefficient of friction and I is the applied torque.

For x near zero, $\sin x \sim x$ and $\cos x \sim 1$, so the equation is approximately linear in x:

$$L\frac{d^2x}{dt^2} + f\frac{dx}{dt} + \left(\frac{d^2V}{dt^2} + g\right)x = I - \frac{d^2H}{dt^2}$$

This linear equation for $x(t)$, whose coefficients vary with t, involves many difficult problems that must be solved to understand the motion of a pendulum. Many of the methods used in the theory of nonlinear oscillations grew out of studies of such pendulum problems; they are applicable now to a wide variety of new problems in physics and biology.

1.2   Time-Invariant Linear Systems

Systems of linear, time-invariant differential equations can be studied in detail. Suppose that the vector of functions $x(t) \in E^N$ satisfies the system of differential equations

$$dx/dt = Ax$$

for $a \le t \le b$, where $A \in E^{N \times N}$ is a matrix of constants.

Systems of this kind occur in many ways. For example, linear nth-order differential equations can be rewritten into the form of a first-order system of equations: Suppose that $y(t)$ is a scalar function that satisfies the linear equation

$$a_n y^{(n)} + \cdots + a_0 y = 0$$

Let us set $x_1 = y, x_2 = y^{(1)}, \cdots, x_{n+1} = y^{(n)}$, and

$$A = \begin{bmatrix} 0 & 1 & 0 & 0 & & \cdots & \\ 0 & 0 & 1 & 0 & & & \\ \vdots & & \ddots & & & \vdots & \\ 0 & 0 & & \cdots & & 0 & 1 \\ b_0 & b_1 & & & & b_{n-2} & b_{n-1} \end{bmatrix}$$

where $b_0 = -a_0/a_n, \cdots, b_{n-1} = -a_{n-1}/a_n$. A matrix in this form is called a *companion matrix* [1.4]. Then the vector x satisfies the differential equation

$$dx/dt = Ax$$

If this were a scalar equation (i.e., $n = 1$), the solution would be $x(t) = \exp(At)x(0)$, where $x(0)$ is given. We show next that this formula also defines a solution when A is any constant matrix.

1.2.1   *Functions of matrices*

Let $g(z)$ be an analytic function of z in some set S of the complex plane. Suppose that g has a convergent power series expansion

$$g(z) = \sum_{n=0}^{\infty} c_n z^n$$

for z in S. If A is an $N \times N$ matrix, then we can make sense of the function $g(A)$ by defining it to be the power series

$$g(\mathbf{A}) = \sum_{n=0}^{\infty} c_n A^n$$

This converges absolutely if the series of scalars $\sum |c_n| \, |A|^n$ converges where $|\cdot|$ denotes the modulus of a complex number and $|\cdot|$ denotes the *Euclidian norm of the matrix A* [1.5]

$$|A| = \sqrt{\sum_i \sum_j |a_{i,j}|}$$

Here $a_{i,j}$ is the component in the ith row and jth column of A.

If A is a diagonalizable matrix, then it can be written in terms of its *spectral decomposition*:

$$A = \sum_{j=1}^{N} \lambda_j P_j$$

where $\lambda_1, \cdots, \lambda_n$ are the eigenvalues of A and $P_1, \cdots, P_n$ are *projection matrices*, which satisfy the conditions $P_i P_j = P_i$ if $i = j$ and $P_i P_j = 0$ otherwise. Because of this, we see that for any integer m,

$$A^m = \sum_{j=1}^{N} \lambda_j^m P_j$$

In addition, we have

$$g(A) = \sum_{n=0}^{\infty} c_n \mathbf{A}^n = \sum_{j=1}^{N} \sum_{n=0}^{\infty} c_n \lambda_j^n P_j = \sum_{j=1}^{N} g(\lambda_j) P_j$$

provided each eigenvalue λ_j lies in the domain where g is analytic.

Note that the spectral decomposition enables us to calculate functions of A in terms of powers of the scalars $\{\lambda_j\}$, rather than powers of A. The result is that once the eigenvalues and their projection matrices are found, effort in calculating functions of A is greatly reduced.

However, not every matrix can be diagonalized. The most that can be said is that any matrix A can be put into *Jordan canonical form*. That is, there is a block diagonal matrix J and a transforming matrix T such that

$$T^{-1}AT = J$$

The blocks on the main diagonal of J have the form

$$\lambda I + Z = \begin{bmatrix} \lambda & 1 & 0 & & 0 & 0 & 0 \\ 0 & \lambda & 1 & \cdots & 0 & 0 & 0 \\ 0 & 0 & \lambda & & 0 & 0 & 0 \\ \vdots & & & \ddots & & \vdots & \\ 0 & 0 & 0 & & \lambda & 1 & 0 \\ 0 & 0 & 0 & \cdots & 0 & \lambda & 1 \\ 0 & 0 & 0 & & 0 & 0 & \lambda \end{bmatrix}$$

where I is an identity matrix of appropriate dimensions ($I_{i,j} = \delta_{i,j}$ and $\delta_{i,j}$ denotes *Kronecker's delta function* ($\delta_{i,j} = 0$ for $i \neq j$, $\delta_{i,j} = 1$ if $i = j$)), and Z is a matrix of zeros except for some 1s on the superdiagonal (where $Z_{i,j} = \delta_{i+1,j}$). Since $Z^N = 0$, Z is referred to as being a *nilpotent matrix*. There may also be diagonalizable terms on the main diagonal of J [see 1.6].

1.2.2 exp(At)

Since $\exp(t)$ is an entire function, $\exp(At)$ can be defined by the series

$$\exp(At) = \sum_{n=0}^{\infty} \frac{(At)^n}{n!}$$

which converges for all (real and complex numbers) t and for any matrix A. We see directly from this series that

$$\frac{d \exp(At)}{dt} = A \exp(At)$$

Therefore, this series defines a solution of the differential equation

$$dx/dt = Ax$$

What does $\exp(At)$ look like? If A is diagonalizable, then we can interpret $\exp(At)$ easily; namely,

$$\exp(At) = \sum_{j=1}^{N} \exp(\lambda_j t) P_j$$

so that $\exp(At)$ is simply a linear combination of matrices with coefficients being exponentials of eigenvalues of A. In general, the series expansion of $\exp(At)$ shows that

$$\exp(At) = \exp(TJT^{-1}t) = T \exp(Jt) T^{-1}$$

Moreover, the exponential of a block diagonal matrix is again a block diagonal matrix having blocks of the same dimensions. Thus, it is sufficient to consider a typical irreducible block matrix:

$$\exp[(\lambda I + Z)t]$$

This has the form

$$\exp(\lambda t)(I + tZ + t^2 Z^2/2 + \cdots + t^k Z^k/k!)$$

where $k + 1$ is the dimension of this block. Therefore, for any matrix A, the function $\exp(At)$ is a linear combination of matrices with coefficients being polynomials in t multiplied by exponentials of eigenvalues of A.

Finally, we note that the spectrum of a matrix A can be split into three parts: Those eigenvalues having negative real parts (S), those having positive real parts (U), and those that are purely imaginary (O). If A is diagonalizable, then it can be transformed into a block diagonal matrix having three blocks: The first consists of those eigenvalues in S, the second of those in O, and the third of those in U. Therefore, we can write

$$\exp(At) = \sum_{j \text{ in } S} \exp(\lambda_j t) P_j + \sum_{j \text{ in } O} \exp(\lambda_j t) P_j + \sum_{j \text{ in } U} \exp(\lambda_j t) P_j$$

The first sum approaches zero as $t \to \infty$, the second one oscillates, and the third one grows as $t \to \infty$. Any solution of the system

$$dx/dt = Ax$$

can therefore be written in the form

$$x(t) = \sum_{j \text{ in } S} \exp(\lambda_j t) P_j x(0) + \sum_{j \text{ in } O} \exp(\lambda_j t) P_j x(0) + \sum_{j \text{ in } U} \exp(\lambda_j t) P_j x(0)$$

In the first sum, $x(0)$ is said to excite *stable modes*; in the second, *oscillatory modes*; and in the third, *unstable modes*. Thus, the matrix A defines a partition of the entire space E^N into three parts: A *stable manifold* that is defined by the span of the projection matrices P_j for $j \in S$, an *unstable manifold* defined by the span of the matrices P_j for $j \in U$, and an oscillatory, or *center manifold*, that is spanned by the matrices P_j for $j \in O$. We see later that this decomposition carries over to certain nonlinear systems.

1.2.3 *Laplace Transforms of Linear Systems*

Suppose that $g(t)$ is a vector of smooth functions that grow no faster than an exponential as $t \to \infty$. We define the *Laplace transform* of g to be

$$g^*(p) = \int_0^\infty \exp(-pt) g(t)\, dt$$

where p is a complex number called the transform variable. Because of our assumption on g, the integrand is dominated by a decreasing exponential if $\text{Re}(p)$ is sufficiently large, where $\text{Re}(p)$ denotes the real part of p. Note that

$$(dg/dt)^* = pg^* - g(0)$$

Therefore, we see that Laplace's transform converts differentiation into multiplication, and it justifies using p as the Heaviside operator described in Section 1.1.2.

How does one recover g from its transform? If $G(p)$ is a function that is analytic in a region except for pole singularities, then we define

$$g(t) = \frac{1}{2\pi i} \int_C \exp(pt)G(p)\,dp$$

where C is a curve lying in the region and enclosing all of the singularities of G. With g defined in this way, $g^*(p) = G(p)$. This formula for g is the *Laplace inversion formula*, and it shows how to recover the original function from its transform.

Calculation of the inverse formula uses the *method of residues*, which is based on *Cauchy's formula*: If $F(z)$ is a function that is analytic in some region containing a point z_0 and if C is a curve lying in this region and enclosing z_0, then we have the formula

$$F(z_0) = \frac{1}{2\pi i} \int_C \frac{F(z)}{(z - z_0)}\,dz$$

This is referred to as the Cauchy integral formula, and the method of residues is based on it. For example, if $G(p) = 1/(p - a)$, then

$$g(t) = \frac{1}{2\pi i} \int_C \frac{\exp(pt)}{(p - a)}\,dp = \exp(at)$$

if C encloses the point $z = a$.

A low-pass filter is described in Section 1.1.2 with $L = 0$. Using the notation of that section, we have

$$V = H(p)W$$

where $H(p) = (RCp + 1)^{-1}$. This formula should be interpreted as one for the transforms of V and W:

$$V^* = H(p)W^*$$

and so $V(t)$ is the inverse Laplace transform of the function $H(p)W^*(p)$. It follows that

$$V(t) = \exp(-t/RC)V(0) + \int_0^t \exp(-(t - s)/RC)W(s)\frac{ds}{RC}$$

Finally, we note that another useful representation of the matrix $\exp(At)$ can be found using Laplace transforms. Namely, we can define

$$\exp(At) = (1/2\pi i) \int_C (pA - I)^{-1}\exp(pt)\,dp$$

This formula is proved by reducing A to its Jordan canonical form and applying Cauchy's formula to each term in the matrix, as shown in the next section.

1.3   Forced Linear Systems with Constant Coefficients

Now let us consider a system of the form

$$dx/dt = Ax + f(t)$$

This describes a linear system to which a forcing function f is applied. We suppose that f grows no faster than an exponential. Taking the Laplace transform of this equation gives

$$(pI - A)x^*(p) = f^*(p)$$

or

$$x^*(p) = (pI - A)^{-1}f^*(p)$$

where this inverse matrix is defined except when p is an eigenvalue of A. Since x^* is the product of two transforms, $x(t)$ is a convolution product; that is, there is a matrix $h(t)$ such that

$$x(t) = \int_0^t h(t - s)f(s)\,ds$$

where $h^*(p) = (pI - A)^{-1}$.

What is h? If A is diagonalizable, then we can use its spectral decomposition to evaluate h. Since

$$A = \sum_{j=1}^N \lambda_j P_j$$

we have that

$$(pI - A)^{-1} = \sum_{j=1}^N \frac{1}{(p - \lambda_j)} P_j$$

Therefore,

$$h(t) = \sum_{j=1}^N \frac{1}{2\pi i} \int_C \frac{\exp(pt)}{(p - \lambda_j)}\,dp\, P_j = \sum_{j=1}^N e^{\lambda_j t} P_j = \exp(At)$$

This motivates the following observation: For any matrix A, we have the function

$$y(t) = \int_0^t \exp[A(t - s)]f(s)\,ds$$

which satisfies

$$dy/dt - Ay = f(t)$$

This formula for y gives a *particular solution* of the equation, and the general solution has the form

$$x(t) = y(t) + z(t)$$

where z solves the homogeneous problem

$$dz/dt - Az = 0 \quad \text{and} \quad z(0) = x(0)$$

so

$$z(t) = \exp(At)x(0)$$

In summary, if $x(t)$ solves the equation

$$dx/dt = Ax + f(t)$$

then

$$x(t) = \exp(At)x(0) + \int_0^t \exp[A(t-s)]f(s)\,ds$$

This is the *variation of constants formula* for $x(t)$, and it might or might not be useful depending on how easily the matrix $\exp(At)$ can be evaluated. The transfer function notation for this is

$$x(t) = (pI - A)^{-1}f(t)$$

so we see that the transfer function notation summarizes a great deal of work. The general variation of constants formula is described in Section 1.6.

1.4   Linear Systems with Periodic Coefficients

Consider the linear system

$$dx/dt = A(t)x$$

where A is now an $N \times N$ matrix of continuous periodic functions, say $A(t + T) = A(t)$ for all t, $-\infty < t < \infty$, where T is a period of A.

We say that $\Phi(t)$ is a *fundamental matrix* for this system (and more generally, any linear system) if

$$\frac{d\Phi}{dt} = A(t)\Phi \quad \text{and} \quad \Phi(0) \text{ is nonsingular}$$

Note that if $\mathbf{d}(t)$ denotes the determinant of $\Phi(t)$, then

$$\frac{d\mathbf{d}}{dt} = \text{tr}[A(t)]\mathbf{d} \quad \text{and} \quad \mathbf{d}(0) \neq 0$$

Therefore,

$$\mathbf{d}(t) = \exp\left\{\int_0^t \text{tr}[A(s)]\,ds\right\}\mathbf{d}(0)$$

so $\mathbf{d}(t) \neq 0$ for all t. It follows that $\Phi(t)$ is nonsingular for all t, and therefore the columns of $\Phi(t)$ define a set of N linearly independent solutions of the system.

The following theorem is very useful for studying periodic systems.

Floquet's Theorem. *Let Φ be as described above. Then there is a periodic matrix $P(t)$, having period T, and a constant matrix R such that*

$$\Phi(t) = P(t)\exp(Rt)$$

Proof of Floquet's Theorem. If $\Phi(t)$ is a fundamental matrix of the problem, then so is $\Phi(t + T)$. Moreover, calculation shows that

$$\frac{d}{dt}\Phi^{-1} = -\Phi^{-1}(t)\Phi'(t)\Phi^{-1}(t)$$

Therefore,

$$\frac{d}{dt}[\Phi^{-1}(t)\Phi(t + T)] = 0.$$

It follows that there is a constant nonsingular matrix, namely

$$C = \Phi^{-1}(0)\Phi(T)$$

such that

$$\Phi(t + T) = \Phi(t)C$$

for all t.

This is the key observation on which further developments are based. In particular, it follows from this formula that for any integer n,

$$\Phi(nT) = \Phi(0)C^n$$

Therefore, the long-term behavior of solutions can be determined from the eigenvalues of the matrix C. If we can define a matrix R by the formula

$$R = \log(C)/T$$

so that $\exp(RT) = C$, then the matrix

$$P(t) = \Phi(t)\exp(-Rt)$$

satisfies the identity

$$P(t + T) = P(t)$$

for all t. In fact, for all t

$$P(t + T) = \Phi(t + T)\exp(-RT)\exp(-Rt) = \Phi(t)\exp(-Rt) = P(t)$$

The logarithm of C is well defined [as shown in 1.7].   ∎

This result is especially helpful in determining the behavior of $x(t)$ as $t \to \infty$. For example, if all of the eigenvalues of R have negative real parts, then $x(t) \to 0$ as $t \to \infty$. However, this theorem is difficult to apply since it is usually difficult to find the matrix R.

An interesting consequence of Floquet's Theorem is that any periodic system can be transformed into one having constant coefficients. In fact, the change of variables $x = P(t)y$ takes the problem

$$dx/dt = A(t)x$$

into the linear system

$$dy/dt = Ry$$

A very useful example is the system

$$\frac{d}{dt}\begin{bmatrix} x_1 \\ x_2 \end{bmatrix} = \begin{bmatrix} \cos \omega t & \sin \omega t \\ -\sin \omega t & \cos \omega t \end{bmatrix} B \begin{bmatrix} \cos \omega t & -\sin \omega t \\ \sin \omega t & \cos \omega t \end{bmatrix} \begin{bmatrix} x_1 \\ x_2 \end{bmatrix}$$

where B is a 2×2 matrix of constants. Using Floquet's transformation,

$$\begin{bmatrix} y_1 \\ y_2 \end{bmatrix} = \begin{bmatrix} \cos \omega t & -\sin \omega t \\ \sin \omega t & \cos \omega t \end{bmatrix} \begin{bmatrix} x_1 \\ x_2 \end{bmatrix}$$

we convert this system into

$$\frac{d}{dt}\begin{bmatrix} y_1 \\ y_2 \end{bmatrix} = \left(B + \omega \begin{bmatrix} 0 & -1 \\ 1 & 0 \end{bmatrix} \right) \begin{bmatrix} y_1 \\ y_2 \end{bmatrix}$$

Thus, in this case, Floquet's transformation is easy to carry out, and $R = B + \omega J$ where J is Jacobi's matrix:

$$J = \begin{pmatrix} 0 & -1 \\ 1 & 0 \end{pmatrix}$$

1.4.1 Hill's Equation

The equation

$$\frac{d^2x}{dt^2} + p(t)x = 0$$

where p is a continuous periodic function, is known as *Hill's equation*. This equaton arises frequently in mathematical physics, for example, Schroedinger's equation in quantum mechanics and studies of the stability of periodic solutions where the linearization about an oscillation often has this form [see 1.3, 1.8].

Let y_1 denote the solution of this equation that satisfies the initial conditions $y_1(0) = 1$ and $y_1'(0) = 0$, and let y_2 denote the solution of this equation that satisfies $y_2(0) = 0$, $y_2'(0) = 1$. Then, if we set $x' = y$, we can rewrite Hill's equation as a first-order system:

$$dx/dt = y$$

$$dy/dt = -p(t)x$$

and the matrix

$$\Phi(t) = \begin{bmatrix} y_1(t) & y_2(t) \\ y_1'(t) & y_2'(t) \end{bmatrix}$$

defines a fundamental solution. Φ is called the *Wronskian matrix* for this system. In fact, each column of this matrix solves the system, and $\Phi(0) = I$.

Suppose that the period p is T, so $p(t + T) = p(t)$ for all t. Floquet's Theorem shows that $\Phi(T) = \exp(RT)$. The eigenvalues of this matrix are called *characteristic multipliers*, and if they have modulus equal to one, then all solutions of Hill's equation are bounded as $t \to \pm\infty$. The eigenvalues of RT are called the *characteristic exponents* of the problem.

On the other hand, a great deal has been determined about the eigenvalues of R for Hill's equation. For example, the eigenvalues of $\Phi(T)$ are determined from the characteristic equation

$$\lambda^2 - [y_1(T) + y_2'(T)]\lambda + \det \Phi(T) = 0$$

and they have the form

$$\lambda = \Delta \pm \sqrt{\{\Delta^2 - \det[\Phi(T)]\}}$$

where $2\Delta = \mathrm{tr}[\Phi(T)]$, the trace of $\Phi(T)$, and det $[\Phi(T)]$ denotes its determinant. Note that $2\Delta = y_1(T) + y_2'(T)$, and det $[\Phi(T)] = \exp\{\int_0^T \mathrm{tr}[A(s)]\,ds\} = 1$.

Therefore, if Δ can be evaluated, then the nature of the eigenvalues of $\Phi(T)$ can be determined. In particular, if $|\Delta| < 1$, then the eigenvalues of $\Phi(T)$ are complex conjugates and have modulus 1. In this case, all solutions of Hill's equation are bounded on the entire interval $-\infty < t < \infty$. On the other hand, if $\Delta > 1$, both roots have positive real parts, and if $\Delta < -1$, then both eigenvalues have negative real parts. In either of these two cases, no solution of Hill's equation remains bounded on the whole interval $-\infty < t < \infty$ [see 1.8]. Finally, if $\Delta = 1$, then there is (at least) one solution having period T, and, if $\Delta = -1$, then there is (at least) one solution having period $2T$. In either case, the other solution can grow no faster than $O(t)$ as $t \to \infty$.

Hill's equation with damping has the form

$$\frac{d^2x}{dt^2} + r\frac{dx}{dt} + p(t)x = 0$$

Setting $x = \exp(-rt/2)y$ gives

$$\frac{d^2y}{dt^2} + \left[p(t) - \left(\frac{r}{2}\right)^2 \right] y = 0$$

This is of the form just considered, and we see that if $\Delta < 1$ for this equation, then y remains bounded for $0 \leq t < +\infty$ and consequently $x \to 0$ as $t \to +\infty$.

Prüfer introduced polar coordinates to Hill's equation setting $y = dx/dt$ and $dy/dt = -p(t)x$. Then $x = r\cos\theta$ and $y = r\sin\theta$. The result is

$$\frac{dr}{dt} = r[1 - p(t)]\cos\theta\sin\theta$$

$$\frac{d\theta}{dt} = -p(t)\cos^2\theta - \sin^2\theta$$

Note that the angular variable θ is separated from the amplitude variable in this case! Thus, Hill's equation is easily put into phase-amplitude variables, which we study further in Section 2.3.5, and reduces to study of the first-order differential equation for θ.

1.4.2 Mathieu's Equation

More can be said about the special case of Hill's equation when $p(t) = \delta + \varepsilon\cos t$ where δ and ε are constants. The result is known as *Mathieu's equation*, and its solutions are either bounded or unbounded, as for Hill's equation. Figure 1.4 shows the values of δ and ε for which solutions are bounded.

Meissner introduced a practice problem, where the term $\cos t$ is replaced by a 2π periodic function $q(t)$ where

$$q(t) = 1 \qquad \text{for } 0 \leq t < \pi$$

$$q(t) = -1 \qquad \text{for } \pi \leq t < 2\pi$$

There is an interesting application of this to the pendulum model described in Section 1.1.4. Let $H'' = 0$ and $V'' = A\cos\omega t$. Then linearizing the pendulum equation about $x = 0$ and replacing t by ωt gives

$$\frac{d^2x}{dt^2} + \left[\left(\frac{g}{L\omega}\right)^2 - \frac{A}{L}\cos t\right]x = 0$$

and linearizing about $x = \pi$ gives

$$\frac{d^2x}{dt^2} - \left[\left(\frac{g}{L\omega}\right)^2 - \frac{A}{L}\cos t\right]x = 0$$

We see that if $\delta = \pm(g/L\omega)^2$ and $\varepsilon = \pm A/L$ lie in the overlap region in Figure 1.5, then both equilibria are stable. Thus, if the support point is vibrated vertically with a range of frequencies and amplitudes so that $(g/L\omega)^2$ and A/L lie in the doubly shaded region in Figure 1.5, then *both* the up and the down positions of the pendulum are stable.

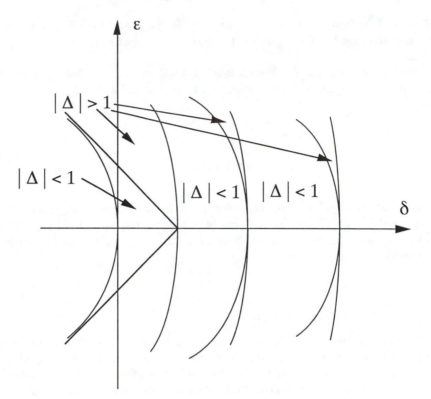

FIGURE 1.4. Stability diagram for Mathieu's equation. If (δ, ε) lies in one of the labeled regions, then $|\Delta| < 1$ and all solutions of Mathieu's equation are bounded. Note that if $\delta = 0$ and $\varepsilon = 0$, then $x(t) = at + b$ for some constants a and b.

If the pendulum is damped, then the linear equations are

$$\frac{d^2x}{dt^2} + r\frac{dx}{dt} \pm \left[\left(\frac{g}{L\omega} \right)^2 - \frac{A}{L}\cos t \right] x = 0$$

respectively. Setting $x = \exp(-rt/2)y$ in each of these equations gives

$$\frac{d^2y}{dt^2} \pm \left[\left(\frac{g}{L\omega} \right)^2 - \frac{A}{L}\cos t \mp \left(\frac{r}{2} \right)^2 \right] y = 0$$

respectively. In these cases, we have that

$$\delta = (g/L\omega)^2 - (r/2)^2 \quad \text{and} \quad \delta = -(g/L\omega)^2 - (r/2)^2$$

respectively.

Figure 1.5 shows the stability diagram for these two oscillators drawn together on the same coordinate axes when $r = 0$. Figure 1.6 shows the same result for small r. In this case small perturbations of the pendulum from the straight up position die out, approximately like $\exp(-rt/2)$. This shows that

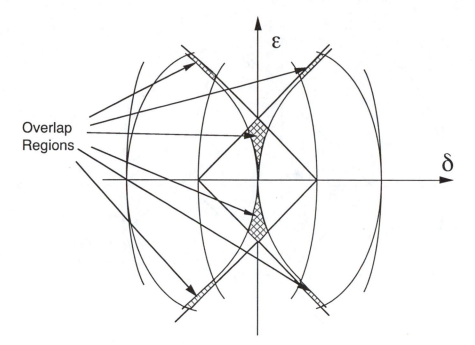

FIGURE 1.5. Overlap region. If the data are in the doubly shaded regions, then both the up and down positions of the pendulum are stable. In this case, a small perturbation of the pendulum from the straight up position persists, but the pendulum remains nearly straight up for all future times.

an oscillating environment can stabilize a static state that is unstable without oscillation. This is an important phenomenon found in many physical and biological systems.

1.5   Fourier Methods

No result comparable to Floquet's Theorem is available for systems having almost periodic coefficients. However, when these sytems do arise in applications, they can be studied using generalized Fourier methods. Some of these methods are described in this section.

1.5.1   *Background on Almost-Periodic Functions*

Almost-periodic functions play important roles in studies of nonlinear oscillators. An almost-periodic function, say $f(t)$, is one that comes close to being periodic in the following sense: for a given tolerance ε there is a number T_ε, called a *translation number*, such that in any interval of length T_ε there is a number T' for which

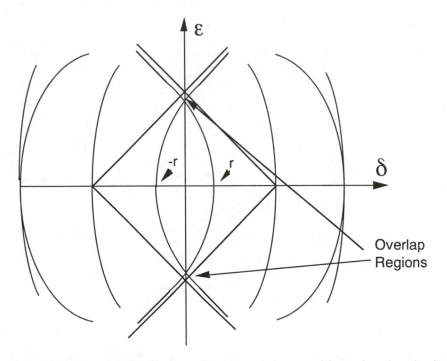

FIGURE 1.6. The stability diagram of the up and down positions when damping is accounted for $(r > 0)$. In this case, small perturbations of the pendulum from the straight up position die out, approximately like $\exp(-rt/2)$.

$$|f(t + T') - f(t)| < \varepsilon$$

for all t. The translation number of a periodic function is simply its period. References 1.9 and 1.10 present introductions to almost-periodic functions and their properties.

If f is *almost periodic*, then it has a generalized Fourier expansion:

$$f(t) \sim \sum_{n=1}^{\infty} C_n \exp(it\lambda_n)$$

where the amplitudes $\{C_n\}$ and the frequencies $\{\lambda_n\}$ characterize the function. The frequencies of f are defined to be the values of λ for which the average

$$A(\lambda) = \lim_{T \to \infty} \frac{1}{T} \int_0^T \exp(-i\lambda t) f(t) \, dt$$

is not zero. It is known that this happens for at most countably many values of λ, say $\lambda_1, \lambda_2, \cdots$. The amplitudes of various modes are given by the formulas

$$C_n = A(\lambda_n)$$

for $n = 1, 2, \cdots$.

An important example of an almost-periodic function is defined by the series

$$f(t) = \sum_{n=1}^{\infty} \frac{\exp(it/2^n)}{n^2}$$

The nth term in the series has period $2^{n+1}\pi$, so the function is not periodic. But, given a tolerance ε, we can choose N so large that

$$\sum_{n=N}^{\infty} \frac{1}{n^2} < \frac{\varepsilon}{2}$$

Then setting $T_\varepsilon = 2^{N+1}\pi$, we have

$$|f(t + T_\varepsilon) - f(t)| < \varepsilon$$

for all t. This shows that f is almost periodic, but note that the integral of this series does not define an almost-periodic function.

The class of all almost-periodic functions is larger than needed to study many nonlinear oscillators, and the smaller class of *quasiperiodic functions* is useful. These functions are generated by finitely many frequencies as follows. Let ω be a vector of M real numbers

$$\omega = (\omega_1, \cdots, \omega_M)$$

and let n be a *multiindex*, that is n is a vector of integers, such that

$$n = (n_1, \cdots, n_M).$$

If the components of ω are rationally related, then the sequence $\{n \cdot \omega\}$ is equivalent to a sequence of integer multiples of a real number. However, if the components of ω are not rationally related, then the sequence is dense in the real numbers. We define $|n| = \sum n_i$, and we consider a set of amplitudes $\{C_n\}$ that satisfies some convergence condition, say $|C_n| \le 1/|n|^2$ as $|n| \to \infty$. Then the series

$$\sum_{|n|=-\infty}^{\infty} C_n \exp(it n \cdot \omega)$$

defines an almost-periodic function whose frequencies are generated by a finite set, namely the components of ω. Such a function is called a *quasiperiodic function*.

Quasiperiodic functions are closely related to periodic functions. For example, if f is a quasiperiodic function, then there is a function $F(s)$, $s \in E^M$, that is periodic in each component of s, such that $f(t) = F(\omega t)$. In particular, if f has the series representation

$$f(t) = \sum_{|n|=-\infty}^{\infty} C_n \exp(it n \cdot \omega)$$

then F is defined by the formula

$$F(s) = \sum_{|n|=-\infty}^{\infty} C_n \exp(in \cdot s)$$

On the other hand, let $F(s_1, \cdots, s_M)$ be a differentiable function that is 2π periodic in each variable $s_1, \cdots, s_M$. Such functions have Fourier series, say

$$F(s_1, \cdots, s_M) = \sum_{|\mathbf{n}|=-\infty}^{\infty} F_{\mathbf{n}} \exp(i\mathbf{n} \cdot \mathbf{s})$$

where $\mathbf{n} = (n_1, \cdots, n_M)$ and $\mathbf{s} = (s_1, \cdots, s_M)$. With a vector of frequencies $\boldsymbol{\omega} = (\omega_1, \cdots, \omega_M)$, we can define a function f by the formula

$$f(t) = F(\omega_1 t, \cdots, \omega_M t)$$

This function has a generalized Fourier series, namely,

$$f(t) = \sum_{|\mathbf{n}|=-\infty}^{\infty} F_{\mathbf{n}} \exp(it\mathbf{n} \cdot \boldsymbol{\omega})$$

so it is a quasiperiodic function.

These remarks show that quasiperiodic functions are essentially multiply periodic functions that comprise more than one frequency.

The following theorem will be used later:

Weak Ergodic Theorem. *Let f be a quasiperiodic function with F defined as above. Suppose that the frequencies defining f, say $\boldsymbol{\omega}$, are rationally independent; that is $\mathbf{n} \cdot \boldsymbol{\omega} \neq 0$ if $\mathbf{n} \neq 0$. Then*

$$\lim_{T \to \infty} \frac{1}{T} \int_0^T f(t) \, dt = \left(\frac{1}{2\pi}\right)^M \int_0^{2\pi} \cdots \int_0^{2\pi} F(s_1, \cdots, s_M) \, ds_1 \cdots ds_M$$

In fact, both sides are equal to C_0, which proves the theorem. This form of the ergodic theorem from statistical mechanics is useful in a variety of numerical computations described in Chapter 7.

1.5.2   *Linear System with Periodic Forcing*

Consider the equation

$$\frac{dx}{dt} = Ax + g(t)$$

where A is a constant diagonalizable matrix and g is a continuous, periodic function; say

$$g(t) = \sum_{m=-\infty}^{\infty} C_m \exp(im\omega t)$$

which has period $2\pi/\omega$. The solution $x(t)$ is given by the formula

$$x(t) = \exp(At)x(0) + \int_0^t \exp[A(t-s)]g(s) \, ds$$

If the spectral decomposition of A is given by

$$A = \sum_{j=1}^{N} \lambda_j P_j$$

then the integral in the formula for $x(t)$ becomes

$$\int_0^t \sum_{j=1}^{N} \sum_{m=-\infty}^{\infty} C_m \exp[\lambda_j(t-s) + i(m\omega)s] \, ds \, P_j = \sum_{j=1}^{N} \sum_{m=-\infty}^{\infty} C_m \frac{\exp(im\omega t)}{(\lambda_j - im\omega)} P_j$$

as long as the ratio $\lambda_j/i\omega$ is not an integer m for which $C_m \neq 0$. This condition ensures that the forcing is *not resonant*.

If the forcing is *resonant*, that is, if $\lambda_j/i\omega = k$ for some integers k and j for which $C_k \neq 0$, then the corresponding term in the solution has the form

$$\int_0^t C_k \exp[\lambda_j(t-s) + ik\omega s] \, ds \, P_j = C_k t \exp(i\omega k t) P_j$$

The resonance case occurs when the forcing function has a frequency that matches one of the free frequencies of the problem, that is, one of the eigenvalues of A. Subresonant forcing occurs when one of the frequencies of g matches an integer multiple of one of the free frequencies.

1.5.3   Linear System with Quasiperiodic Forcing

Now, consider the equation

$$\frac{dx}{dt} = Ax + g(t)$$

where A is a constant diagonalizable matrix and g is a continuous, quasiperiodic function generated by two frequencies, say ω and μ:

$$g(t) = \sum_{m,n=-\infty}^{\infty} C_{m,n} \exp[i(m\omega + n\mu)t]$$

We suppose that $|C_{m,n}| \leq K/(m^2 + n^2)$ for some constant K.

The solution $x(t)$ is given by the formula

$$x(t) = \exp(At)x(0) + \int_0^t \exp[A(t-s)]g(s) \, ds$$

If the spectral decomposition of A is

$$A = \sum_{j=1}^{N} \lambda_j P_j$$

then the integral in the formula for $x(t)$ becomes

$$\int_0^t \sum_{j=1}^{N} \sum_{m,n=-\infty}^{\infty} C_{m,n} \exp[\lambda_j(t-s) + i(m\omega + n\mu)s] \, ds$$

$$= \sum_{j=1}^{N} \sum_{m,n=-\infty}^{\infty} C_{m,n} \frac{\exp[i(m\omega + n\mu)t]}{i(m\omega + n\mu) - \lambda_j} P_j - \sum_{j=1}^{N} \sum_{m,n=-\infty}^{\infty} \frac{C_{m,n}\exp(\lambda_j t)}{i(m\omega + n\mu) - \lambda_j} P_j$$

if these series converge. The series do converge if the eigenvalues of A are not purely imaginary. However, if some eigenvalues of A lie on the imaginary axis and the frequencies μ and ω are rationally independent, then the series representing $x(t)$ becomes more difficult to evaluate. In particular, the numbers $i\{m\omega + n\mu\}$ are dense in the imaginary axis, and so the denominators $i(m\omega + n\mu) - \lambda_j$ become arbitrarily small for large m and n. This is a *small divisor problem*, and we will deal with aspects of it in Sections 7.4.3 and 7.5.3.

1.6   Linear Systems with Variable Coefficients: Variation of Constants Formula

A general linear system has the form

$$\frac{dx}{dt} = A(t)x + f(t)$$

where x, $f \in E^N$ and A is an $N \times N$-dimensional matrix of continuous functions. Let $\Phi(t)$ denote the solution of the matrix equation

$$\frac{d\Phi}{dt} = A(t)\Phi \qquad \Phi(0) = \text{identity}$$

Φ is called the *fundamental solution* for the problem, and it is determined by solving these N^2 scalar differential equations for the components of Φ.

The solution $x(t)$ of the linear problem is given by the *variation of constants* formula

$$x(t) = \Phi(t)x(0) + \int_0^t \Phi(t)\Phi^{-1}(s)f(s)\,ds$$

This formula can be easily derived by setting $x = \Phi(t)y$ since the equation for y is

$$\frac{dy}{dt} = \Phi^{-1}(t)f(t)$$

The difficulty in using this formula is that Φ is usually not available. If A is constant, then we have seen that

$$\Phi(t) = \exp(At)$$

If A is periodic, then

$$\Phi(t) = P(t)\exp(Rt)$$

which at least shows the form of Φ even though P and R may be difficult to find. No comparable result is available in cases where A is quasiperiodic or almost periodic. In the general case, note that usually

$$\Phi(t) \neq \exp\left(\int_0^t A(s)\, ds\right),$$

since in general

$$A(t)\int_0^t A(s)\, ds \neq \int_0^t A(s)\, ds\, A(t)$$

1.7 *Exercises*

1.1. a. Show that the Laplace transform of a convolution

$$\int_0^t h(t-s)f(s)\, ds$$

is the product of the transforms of h and f if f and h do not grow faster than exponential functions as $t \to \infty$. Evaluate the Laplace transform of a derivative dg/dt of a differentiable function $g(t)$.

b. Use the Laplace transform to solve the differential equation

$$\frac{d^2x}{dt^2} + \frac{dx}{dt} + bx = f(t)$$

where a and b are known constants and $f(t)$ is a given function.

c. Show that if

$$y(t) = \int_0^t \exp[A(t-s)]f(s)\, ds$$

where y, f in E^N and A in $E^{N \times N}$ is a constant matrix, then

$$dy/dt - Ay = f$$

Suppose that A is diagonalizable. Use the spectral decomposition of A and the Laplace transform to relate the solution of this differential equation to the eigenvalues of A.

d. Show that any solution of the differential equation

$$dy/dt - Ay = f$$

can be written in the form

$$y(t) = \exp(At)y(0) + \int_0^t \exp[A(t-s)]f(s)\, ds$$

using the method of Laplace transforms.

1.2. a. Let $P(t) = \Phi(t)\exp(-Rt)$ as in the proof of Floquet's Theorem in Section 1.4. Show that the change of variables $x = P(t)y$ takes the periodic system

$$dx/dt = A(t)x$$

into the system

$$dy/dt = Ry$$

where R is a constant matrix. Thus, show that any periodic system can be transformed in this way into a linear system having constant coefficients.

b. Find a 2×2 matrix of functions, say $A(t)$, such that

$$A(t)\left[\int_0^t A(s)\,ds\right] \neq \left[\int_0^t A(s)\,ds\right]A(t)$$

Show that in this case the matrix $\Phi(t) = \exp[\int_0^t A(s)\,ds]$ does not solve the equation

$$d\Phi/dt = A(t)\Phi$$

Find such a matrix $A(t)$ that is periodic. Relate this result to Floquet's theorem.

c. Show that if $\Phi(t)$ is an $N \times N$ matrix of functions satisfying the equation

$$d\Phi/dt = A(t)\Phi$$

then the function $D(t) = \det[\Phi(t)]$ satisfies the equation

$$dD/dt = \mathrm{tr}[A(t)]D(t)$$

where $\mathrm{tr}[A]$ is the trace of A (i.e., the sum of the elements of A lying on the main diagonal.)

1.3. Let $A(t) = U^{-1}(t)CU(t)$ where C is a constant 2×2-matrix and the components of the 2×2-matrix U are

$$U_{11} = \cos t, \qquad U_{12} = \sin t, \qquad U_{21} = -\sin t, \qquad U_{22} = \cos t.$$

The eigenvalues of A and C are identical. Apply Floquet's transformation to this system. Show by example that the resulting system having constant coefficients can have a positive eigenvalue even though the eigenvalues of C both have negative real parts. Conclude that eigenvalues of a periodic coefficient matrix do not determine stability properties of the linear system.

1.4. Show that if $|D| < 1$, then all solutions of Hill's equation (Section 1.4.1.) are bounded for $-\infty < t < \infty$. Also, show that if $|D| > 1$, then all solutions (excluding the one that is zero everywhere) are unbounded on $-\infty < t < \infty$.

1.5. Construct the complete stability diagram for Meissner's equation in Section 1.4.2. using numerical simulation.

1.6. Verify the weak ergodic theorem for the function $F(s_1, s_2) = 1 + \cos s_1 + \cos s_2$ and $\omega_1 = 1$ and $\omega_2 = \sqrt{2}$ by direct substitution. That is, show that

$$(2\pi)^{-2}\int_0^{2\pi} F(s_1, s_2)\,ds_1\,ds_2 = \lim_{T\to\infty} \frac{1}{T}\int_0^T F(t, \sqrt{2}t)\,dt$$

2
Free Oscillations

A distinction is usually made between systems that are isolated, known as *free systems,* and those that interact with the outside world, known as *forced systems.* Often we reduce forced systems to (apparently) free ones by looking at the system stroboscopically or by introducing extra variables to describe external influences. Often we reduce free problems to ones that appear to be forced. For example, systems in which energy is conserved can have dissipative components within them, which can be uncovered by finding a timelike variable among the variables of the system and using it to reduce the problem to a dissipative one of lower order.

The term *oscillation* usually refers to a periodic or almost-periodic solution of a system of differential equations. *Free oscillations* are oscillatory solutions to models that are time invariant. The LC circuit is a typical free oscillator. *Nonlinear oscillations* are oscillatory solutions to nonlinear differential equations. This chapter begins with a study of free nonlinear problems in two variables for which a great deal is known.

Most oscillations studied here can be described by functions of the form $F(x_1, \cdots, x_N)$ that are 2π periodic in each of N phase variables $x_1, \cdots, x_N$, which themselves vary with time. A convenient mathematical description of an oscillator is one given in terms of such phase variables, and phase equations are studied in Section 2.2. However, most physical problems are described in terms of physical observables like voltages, currents, displacements, and so on, and converting such problems to convenient phase variables is usually difficult. The third and fourth sections investigate general conservative and dissipative systems, respectively. This leads to a discussion of discrete dynamics and function iterations in Section 2.5. Finally, some oscillations caused by time delays are described in Section 2.6.

2.1 Systems of Two Equations

It is possible to describe in detail solutions of linear equations for two variables. Much of this work carries over to nonlinear systems as well, at least near equilibria. Poincare and Bendixson's theory deals with general nonlinear systems for two variables. It is derived in Section 2.1.2 and then later applied to study Lienard's equation.

2.1.1 *Linear systems*

Consider the linear system

$$\frac{dx}{dt} = ax + by$$

$$\frac{dy}{dt} = cx + dy$$

where a, b, c, and d are real numbers. The point $x = 0$, $y = 0$, is an equilibrium for this system, and the following analysis shows how all solutions of the system behave.

We saw in Chapter 1 that the solutions of this system are based on exponentials of eigenvalues of the coefficient matrix, namely, solutions of the characteristic equation

$$\lambda^2 - (a + d)\lambda + (ad - bc) = 0$$

where the coefficient of $-\lambda$ is tr(A), the trace of the matrix

$$A = \begin{bmatrix} a & b \\ c & d \end{bmatrix}$$

and the last term is its determinant, det A. Therefore this equation has the form

$$\lambda^2 - \text{tr}(A)\lambda + \det(A) = 0$$

The eigenvalues are

$$\lambda = \frac{\text{tr}(A)}{2} \pm \sqrt{\left(\frac{\text{tr}(A)}{2}\right)^2 - \det(A)}$$

If we suppose that det $A \neq 0$, then there are three cases:

1. The roots are real and have the same sign. If the roots are negative, then all solutions tend to the critical point as $t \to \infty$. This is called a *stable node* (see Figure 2.1). In the other case, it is an *unstable node*.
2. The roots are real and have opposite signs. In this case, there is a one-dimensional stable manifold and a one-dimensional unstable manifold (see Section 1.2.2). The critical point $x = 0$, $y = 0$ is called a *saddle point* or *col* in this case (see Figure 2.2).
3. The roots are complex conjugates. If the real part of the eigenvalues is negative, then all solutions spiral into the critical point. This is a *stable spiral* or a *spiral sink* (Figure 2.3). If the real part is positive, solutions spiral away from the critical point. This is an *unstable spiral* or a *spiral*

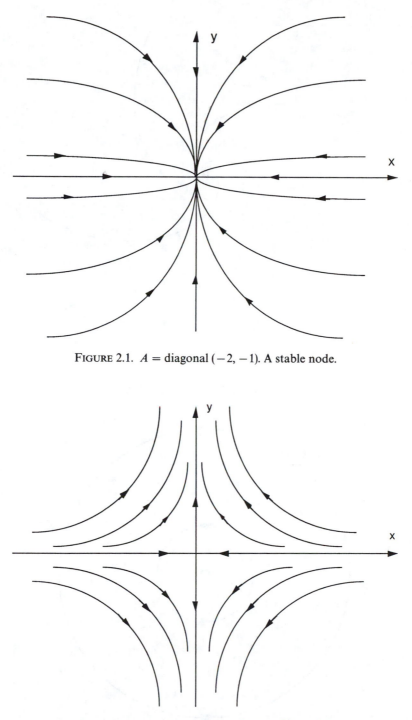

FIGURE 2.1. A = diagonal $(-2, -1)$. A stable node.

FIGURE 2.2. A = diagonal $(1, -1)$. A saddle point.

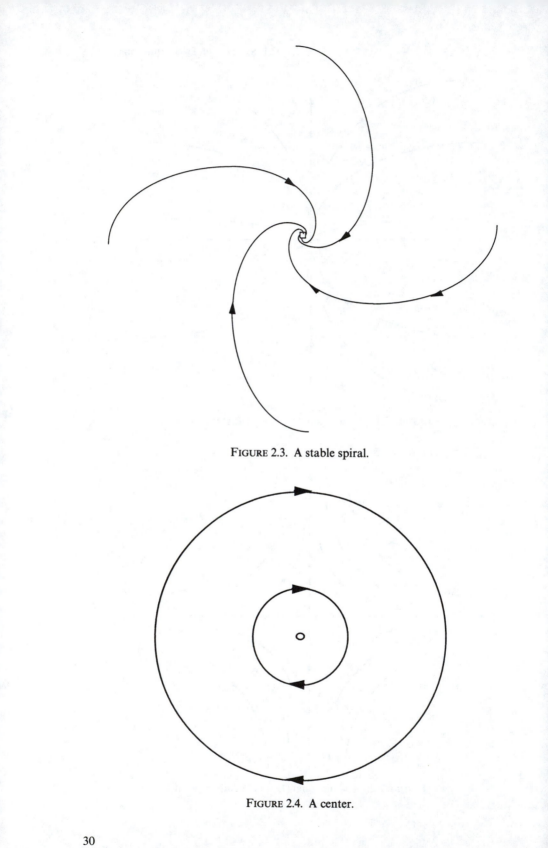

FIGURE 2.3. A stable spiral.

FIGURE 2.4. A center.

source. If the real parts are zero, then this is a *center*, and solutions circle around it (see Figure 2.4).

Much of the behavior depicted in Figures 2.1–2.4 is also seen in nonlinear systems. However, nonlinear systems can have more complicated behavior.

2.1.2 Poincare and Bendixson's Theory

Consider a system of two ordinary differential equations of the form

$$\frac{dx}{dt} = f(x, y)$$

$$\frac{dy}{dt} = g(x, y)$$

where f and g are continuously differentiable functions of (x, y) over some compact (i.e., a closed and bounded) set D in E^2. Given a point (ξ, η) in D there is a unique solution of this system passing through it. We say that (ξ, η) is an equilibrium (equivalently, rest point, critical point, or static state) for the system if $f(\xi, \eta) = g(\xi, \eta) = 0$. Otherwise, we write the unique solution beginning $(t = 0)$ at (ξ, η) as $x = x(t, \xi, \eta)$, $y = y(t, \xi, \eta)$. Poincare and Bendixson's theory describes what can happen to this solution if it eventually remains in D.

Poincare-Bendixson Theorem. *Suppose that D is a closed and bounded subset of E^2 containing at most a finite number of rest points. Moreover, suppose that $(x(t), y(t))$ is a solution of the system*

$$\frac{dx}{dt} = f(x, y)$$

$$\frac{dy}{dt} = g(x, y)$$

and that for some time t_0 it remains in D for all $t \geq t_0$. Then there are three possibilities:

1. *$[x(t), y(t)]$ tends to a critical point as $t \to \infty$.*
2. *$[x(t), y(t)]$ tends to a periodic orbit as $t \to \infty$. This means that there is a periodic solution of the system, say $x = p(t)$ and $y = q(t)$ such that the solution $[x(t), y(t)]$ approaches the set $\{[p(\tau), q(\tau)]: -\infty < \tau < \infty\}$ as $t \to \infty$.*
3. *$[x(t), y(t)]$ approaches a set that is homeomorphic to a circle and that contains critical points and orbits joining them.*

Thus, either solutions approach a critical point of the system, an oscillation, or they approach a necklace of critical points on a strand of orbits.

The proof of this result can be found in [Refs. 2.1–2.3], and it is not presented here.

This remarkable result rules out a great number of pathologies for two-dimensional systems that must be dealt with in higher dimensional systems.

Much of the work on differential equations is restricted to problems that can be reduced to systems to which this theorem can be applied. A typical application of the Poincare-Bendixson theory is given in Section 2.1.3.

This theorem is not valid in dimensions higher than two. It relies essentially on reducing the system to a single scalar ordinary differential

$$\frac{dx}{dy} = \frac{f(x, y)}{g(x, y)}$$

and then using special properties of solutions to such equations. *Lorenz's system* of equations provides an important example of three equations where most solutions do not converge to any of the three options in the Poincare-Bendixson Theorem. Instead, they wander in a structured but chaotic way for all time. Lorenz's system is

$$\frac{dx}{dt} = \sigma(y - x)$$

$$\frac{dy}{dt} = xz + \alpha x - y$$

$$\frac{dz}{dt} = xy - \beta z$$

where α, β, and σ are constants. Chaotic solutions of this system are present when $\sigma = 10$, $\alpha = 28$, and $\beta = 8/3$.

2.1.3 $x'' + f(x)x' + g(x) = 0$

Lienard's equation includes a variety of nonlinear second-order equations that are used in many applications. It includes the general form

$$\frac{d^2x}{dt^2} + f(x)\frac{dx}{dt} + g(x) = 0$$

where f and g are continuous functions of x. Two cases of special importance are *van der Pol's equation*

$$f(x) = ax^2 - b \qquad \text{and} \qquad g(x) = \omega^2 x$$

and *Duffing's equation*

$$f(x) = 0 \qquad \text{and} \qquad g(x) = x^3 + ax + b$$

for some constants a and b. These two cases are studied at several points throughout this book. However, in this section we study a form of Lienard's equaton that includes van der Pol's equation, namely,

$$\frac{d^2x}{dt^2} + f(x)\frac{dx}{dt} + x = 0$$

where f is a smooth function. We now derive conditions that ensure this equation supports a nonlinear oscillation.

In analogy with mechanical systems, we define the energy of the system to be the sum of the kinetic and potential energies:

$$\frac{\dot{x}^2}{2} + \frac{x^2}{2}$$

In this case we have

$$\frac{\dot{x}^2}{2} + \frac{x^2}{2} = C - \int_0^t f[x(s)]\dot{x}^2(s)\,ds$$

where C is a constant. Thus, if $f = 0$, energy is conserved; if $f(x) > 0$, then energy is taken out of the system; and, if $f(x) < 0$, then energy is put into the system.

Integrating the differential equation once results in an equivalent first-order system:

$$dx/dt = y - F(x)$$
$$dy/dt = -x$$

where

$$F(x) = \int_0^x f(z)\,dz$$

Hypothesis H.

1. $F(-x) = -F(x)$ for all x.
2. There is a number $\alpha > 0$ such that $F(x)$ is negative for $0 < x < \alpha$.
3. There is a number $\beta \geq \alpha$ such that $F(x)$ is positive and strictly increasing for $x > \beta$.
4. $F(\infty) = \infty$.

In analogy with the *RLC* filter (see Section 1.1.2), the coefficient $f(x)$ corresponds to a resistance: If $f > 0$, then there is resistance and energy is being taken out of the system. If $f < 0$, then there is negative resistance, and energy is being put into it. The fact that $F < 0$ for x near 0 implies that $x = 0$ lies in a negative resistance region. The fact that $F > 0$ for large x indicates that the system is resistive for large x. We see examples of electronic devices, such as tunnel diodes, that exhibit both positive and negative resistances later. With Hypothesis H, we have the following

Lienard's Theorem. *Let Hypothesis H be satisfied. Then Lienard's equation has a (nontrivial) periodic solution.*

Proof. Consider a solution of the system starting with $x(0) = 0$ and $y(0) = y_0 > 0$. If y_0 is large, then there is a solution of the equation that proceeds clockwise completely around the origin and hits the positive y axis at a point $y_1(y_0)$.

We use the following lemma, which we state here without proof (see Ref. 2.2, pp 57–60).

Lemma. The mapping $y_1: y_0 \rightarrow y_1(y_0)$ has the following properties:

1. If $\alpha < \beta$, then y_1 is defined for large values of y_0 and for small values of y_0. In addition, for large values of y_0, $y_1(y_0) < y_0$, and for small values of y_0, $y_1(y_0) > y_0$.
2. If $\alpha = \beta$, then y_1 is defined for all values of y_0 on the positive y axis. Moreover, y_1 is a monotone decreasing function of y_0.

The results in the above lemma (1) are illustrated in Figure 2.5. Since there are no critical points in the annular region described in Figure 2.5, it follows

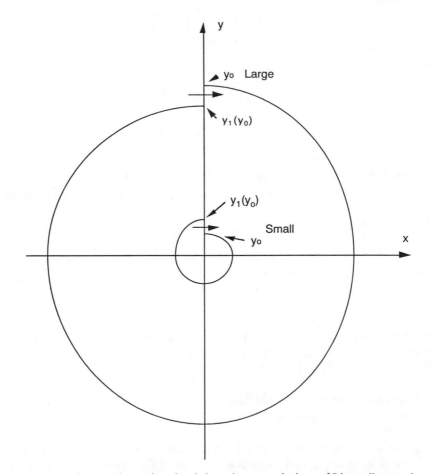

FIGURE 2.5. An annular region that is invariant to solutions of Lienard's equation.

from the Poincare-Bendixson Theorem that any solution starting on the (upper) line segment $[y_1(y_0), y_0]$ must approach a periodic solution of the system. This completes the proof of the theorem.

An interesting corollary of this theorem uses the additional condition:

Hypothesis H'. $\alpha = \beta$
With this we have the following theorem.

Theorem. *Let conditions H and H' be satisfied. Then Lienard's equation has a unique periodic solution. All solutions in the plane approach this orbit except for the unstable equilibrium $x = 0$, $y = 0$.*

Proof. The proof results from the above lemma (2). With condition H' the mapping $y_0 \to y_1(y_0)$ is monotone decreasing. Since the previous theorem shows this mapping has a fixed point, it must be unique. This completes the proof [see also Ref. 2.1]. ∎

Example of van der Pol's Equation. We can use these results to study van der Pol's equation

$$\frac{d^2x}{dt^2} + A(x^2 - 1)\frac{dx}{dt} + x = 0$$

In this case, $F(x) = A(x^3/3 - x)$ and $\alpha = \beta = \sqrt{3}$. The last theorem shows that there is a unique periodic solution of van der Pol's equation for any choice of the parameter A ($A > 0$). Therefore, van der Pol's equation has a globally stable periodic solution for any $A > 0$. We study this nonlinear oscillation when $A \ll 1$ and when $A \gg 1$ in Chapters 7 and 8, respectively.

2.2   Angular Phase Equations

Unfortunately, the term *phase equation* is used here with at least two different meanings. First, phase refers to an angular variable, as in the phase of the moon; second, phase variables are position and momentum variables in mechanical systems. In this section, we study angular phase variables.

Phase equations give a convenient description of nonlinear oscillators. In the next section, we show how some other models can be transformed into phase-amplitude coordinates, but in this section we consider systems of (angular) phase equations. In general, these models are measured by an observable, say $F(x_1, \cdots, x_N)$, which is 2π periodic in each of the phases, $x_1, \cdots, x_N$. Because of this, we need only consider these variables in a set

$$T^N = \{(x_1, \cdots, x_N): 0 \le x_j < 2\pi \text{ for } j = 1, \cdots, N\}$$

T^N is referred to as the N torus. For $N = 1$, it can be identified with a circle where x is the angular coordinate of the circle (Figure 2.6); for $N = 2$, it can be identified with a torus that is constructed by cutting a patch out of the

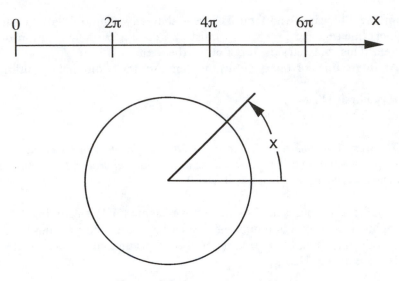

FIGURE 2.6. An angular variable.

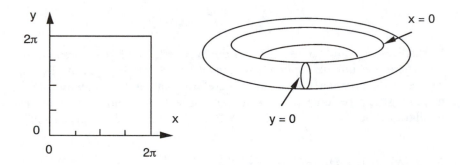

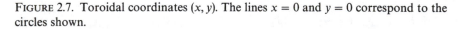

FIGURE 2.7. Toroidal coordinates (x, y). The lines $x = 0$ and $y = 0$ correspond to the circles shown.

$x_1 x_2$ plane that is length 2π on a side and identifying the edge $x_1 = 0$ with the edge $x_1 = 2\pi$ by sewing them together to form a cylinder, and then identifying the edge $x_2 = 0$ with the edge $x_2 = 2\pi$ by sewing them together (Figure 2.7). The general case is more difficult to visualize, although one can usefully think of T^N being a cube cut out of E^N with opposite sides identified.

Where do phase equations come from? First, we see in Section 2.3.5 that phase-amplitude coordinates can be used effectively to study some nonlinear oscillators. Two important mechanical cases are perturbed harmonic oscillators and weakly coupled pendula where phase variables describe anuglar momentum. Second, modern electronic circuits are designed to be modeled directly using phase variables to facilitate their use in more complicated circuits. This is illustrated by the VCO in Chapter 1.

2.2.1   A Simple Clock. A Phase Equation on T^1

A single phase is an angular variable describing a circle as shown in Figure 2.6:

Let us consider a clock having one hand with its circumference being numbered from 0 to 2π (counterclockwise). The hand moves at a constant rate of ω radians per hour, so if x denotes the angle between the hand and 0, then x satisfies the differential equation

$$dx/dt = \omega$$

where t is measured in hours. The solution of this equation is $x(t) = \omega t + x_0$.

Scalar equations of the form

$$dx/dt = f(x)$$

cannot have periodic solutions. In fact, if $x(t)$ is an oscillatory solution of this equation, then there must be a time when $dx/dt = 0$, so the solution must have hit an equilibrium. But, since solutions are unique, $x(t)$ must be an equilibrium. Also, solutions $x(t)$ are monotone between equilibria.

On the other hand, the VCO in Chapter 1 is modeled by a simple phase equation, and for it an oscillation means that $x \to \infty$ since the observed voltage, $\cos x$, is a periodic function of x. We say that there is a periodic solution for x on the circle. (i.e., x modulo 2π)

The following circuit is a slight extension of the VCO:

$$\frac{dx}{dt} = \omega_0 + E + V(x)$$

This describes a feedback circuit in which the voltage put out by the VCO is augmented by an acquisition voltage E and fed back as the control for the VCO. The phase of the voltage put out by the VCO, say x, satisfies the equation

$$\frac{dx}{dt} = \omega + f(x)$$

where $f(x)$ is $V(x) - \langle V \rangle$, the voltage put out by the VCO less its average value, and ω is the sum of the center frequency ω_0, the acquisition voltage E, and the average $\langle V \rangle$. We have put the mean value of $V(x)$ into ω, so f is a periodic function having mean zero.

There are two cases to be studied:

1. $|\omega| \le \max |f(x)| \equiv F$. This implies that there is a static state, and since the solutions of this equation are monotone between static states, each tends toward a static state as $t \to \infty$. In particular, for any initial condition, $x(t) \to$ constant as $t \to \infty$. In this case, $V(x) \to$ constant.
2. $|\omega| > F$. This implies that $|x| \to \infty$ as $t \to \infty$, and so there is a periodic solution. The equation can be solved by quadrature using the expression

$$dt = \frac{dx}{\omega + f(x)}$$

Since f has period 2π in x, we have that t increases by T units when x increases by 2π units where

$$T = \int_0^{2\pi} \frac{dx}{\omega + f(x)}$$

In the case of the feedback VCO circuit [with $f(x) = \cos x$]

$$T = \tan^{-1}\left[\frac{\sqrt{(\omega_0 + E)^2 - 1}\sin x}{1 + (\omega_0 + E)\cos x}\right]\bigg/\sqrt{(\omega_0 + E)^2 - 1}$$

Thus, a trigonometric polynomial results for x.

It is plausible that $x \sim 2\pi t/T$ for large t. If so, the VCO output is eventually oscillatory with frequency $\omega_1 = 2\pi/T$.

Another important example of a single phase equation is the harmonic oscillator. Consider the equation

$$u'' + \omega^2 u = 0$$

If we introduce polar coordinates to this equation by setting

$$u' + i\omega u = ye^{ix}$$

then

$$y' = 0 \qquad \text{and} \qquad x' = \omega$$

which is the phase-amplitude version of a harmonic oscillator.

2.2.2   A Toroidal Clock: Denjoy's Theory

Two simple clocks, say with phases x and y, respectively, are modeled by the equations

$$dx/dt = \omega$$

$$dy/dt = \mu$$

Telling time on this pair of clocks is like locating a point on T^2, so we refer to this as a *toroidal clock*. For example, let t be solar time in days, let x be the phase of a solar day, and let y be the phase of the moon. Then $\omega = 2\pi$ and $\mu = 2\pi/T$, since the x variable must complete a cycle of length 2π each day and the y variable must complete a cycle of 2π every T days (approximately 29 days).

A general pair of phase equations has the form

$$dx/dt = f(x, y)$$

$$dy/dt = g(x, y)$$

where f and g are smooth functions that are 2π periodic in each of x and y.

Since only values of x and y modulo 2π play a role, we might as well consider this system to be on the xy plane reduced modulo 2π. As noted earlier, this is equivalent to a torus T^2. Therefore, this doubly periodic system is often referred to as being a system of ordinary differential equations on a torus, which is generated by the two circles labeled $x = 0$ and $y = 0$ in Figure 2.7.

Poincare and Denjoy studied this problem in some detail [see 2.1]. They defined the *rotation number* to be the limit

$$\rho = \lim_{t \to \infty} \frac{x(t)}{y(t)}$$

The limit ρ gives the relative frequency between x and y, and it has some very interesting and useful features. Denjoy derived the following results [2.4].

Denjoy's Theorem. *Let f, g be as above. Then the number ρ has the following properties*:

1. *ρ exists, and its value is independent of the initial values, $x(0)$ and $y(0)$, used to compute it.*
2. *ρ depends continuously on f and g.*
3. *If ρ is rational, then every solution of this system approaches a periodic solution on the torus, called a torus knot. Moreover, if $\rho = p/q$ where p and q are integers, then the solution rotates in the x direction p times around the torus and in the y direction q times around the torus each period.*
4. *If ρ is irrational, then all solutions of the system are dense in the torus. This is referred to as the ergodic case.*

The rotation number is usually difficult to evaluate, but it can be simulated using a computer, and in that way periodic solutions on the torus can be detected by simulation.

Figure 2.8 shows a simulation of the rotation number for the system

$$\frac{dx}{dt} = \omega + \lambda \sin(x - y) + (1 - \lambda)[\sin(2x - y) - \sin(2y - x)]$$

$$\frac{dy}{dt} = \mu$$

Denjoy's theory leads to a good numerical method for describing the behavior of nonlinear oscillators. We will use the theory again in Chapter 7.

2.2.3   Systems of N (Angular) Phase Equations

Systems of angular phase equations have the form

$$\frac{dx}{dt} = \omega + f(x)$$

where $x, f, \omega \in E^N$, and f is a vector of functions that are 2π periodic in each

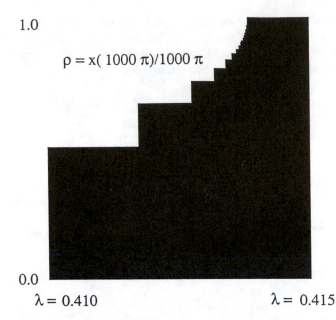

1.0

$\rho = x(\,1000\,\pi)/1000\,\pi$

0.0

$\lambda = 0.410$ $\lambda = 0.415$

FIGURE 2.8. Rotation number simulation for $x' = \lambda \sin(x - t) + (1 - \lambda)\sin(2x - t)$. While the transitions between 1/2 and 2/3, etc., seem to be discontinuous, Denjoy's theory ensures that this function is a continuous function of λ. Further simulation of the rotation number at these "jumps" shows many steps. The function is similar to Cantor's function.

component of x. Without loss of generality we assume that

$$\frac{1}{2\pi} \int_0^{2\pi} f(x)\,dx = 0$$

These equations arise in celestial mechanics, statistical mechanics, modern circuit theory, and many other applications. Not much can be said about them in general, although if $\omega_j \gg f_j$, then averaging methods can be used, as shown in Chapter 7.

We also see in Chapter 7 that the rotation vector method extends Poincaré's theory of rotation numbers to higher dimensions.

2.2.4   *Equations on a Cylinder: VCON.*

The circuit in Figure 2.9 is a phase-locked loop [2.5]. A voltage W_+ comes into a phase detector whose output is $V_+(x)W_+$. This signal passes through a low-pass filter, and the result controls the VCO. The mathematical formulation of a phase-locked loop is as follows: The unknowns in this model are the filter output voltage $z(t)$ and the VCO phase $x(t)$. The wave form V is fixed, and in commercially available VCOs it can be a square wave, a triangular wave, or a sinusoidal wave. $V(x)$ is a 2π-periodic function of the phase x. The

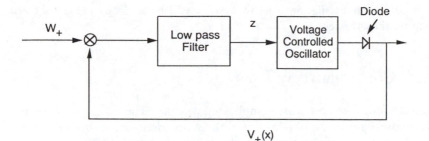

FIGURE 2.9. A voltage-controlled oscillator neuron (VCON) model.

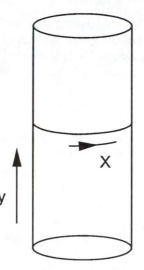

FIGURE 2.10. Cylindrical coordinates.

model for this circuit is

$$dx/dt = \omega + z$$

$$\tau \, dz/dt = -z + V_+(x)W_+$$

The first equation describes the VCO: ω is the VCO center frequency and z is its controlling voltage. The second equation describes the output z of the low-pass filter and τ gives the time constant of this filter. This model is quite similar to a pendulum with an oscillating support point and an applied torque [2.7] and to a point Josephson junction [2.8].

If W is constant, this model has one phase variable x and one voltage variable z. It is convenient to visualize this model geometrically as being one on a cylinder as shown in Figure 2.10. This model is studied further in Section 2.4.2.

A nonlinear oscillation for this system will be a closed curve that goes around the cylinder.

2.3   Conservative Systems

Second-order oscillators are complicated to study. The easiest ones are those derived from conservation laws, and these are discussed in this section. But, first we recall some notations and ideas from mechanics [2.9].

2.3.1   *Lagrangian Mechanics*

Consider a mechanical system whose position or state at time t is described by a scalar function $x(t)$. Let the potential energy of this system be denoted by $U(x)$, and denote the kinetic energy by $m\dot{x}^2/2$, where m is mass. The Lagrangian of this system is defined to be

$$L = m\frac{\dot{x}^2}{2} - U(x)$$

and Hamilton's principle states that the system moves between any two times a and b in such a way as to make the action integral

$$I(x) = \int_a^b L(x, \dot{x})\, dt$$

stationary. That is, if we perturb x by a continuously differentiable function $\varepsilon\eta(t)$ where $\eta(a) = \eta(b) = 0$ and where ε is a small positive real number, then we consider the integral

$$I(x + \varepsilon\eta) = I(x) + \varepsilon I'(x)\eta + O(\varepsilon^2)$$

where

$$I'(x)\eta = \int_a^b \left(\frac{\partial L}{\partial x}\eta + \frac{\partial L}{\partial \dot{x}}\dot{\eta}\right) dt$$

The integral is *stationary* if $I'(x)\eta = 0$ for any function η as described above.

A necessary condition for this is obtained by integrating this formula by parts and setting the resulting integrand equal to zero:

$$\frac{\partial L}{\partial x} - \frac{\partial}{\partial t}\left(\frac{\partial L}{\partial \dot{x}}\right) = 0$$

This is the *Euler-Lagrange* equation for $I(x)$ [2.9]. The result in this case is that

$$m\frac{d^2x}{dt^2} + \frac{\partial U}{\partial x} = 0$$

Multiplying both sides of this equation by $\dot{x}$ and integrating it gives

$$\frac{m}{2}\left(\frac{dx}{dt}\right)^2 + U(x) = \text{constant}$$

Thus, the total energy of the system is conserved.

2.3.2 Plotting Phase Portraits Using Potential Energy

Consider the conservation equation

$$\frac{d^2x}{dt^2} + \frac{\partial U}{\partial x} = 0$$

where the potential function U is a smooth function of x. Maxima of U correspond to saddle points and minima correspond to centers in the phase portrait of solutions to this equation. To see this, suppose that x^* is an extremum of U; that is, $(\partial U/\partial x)(x^*) = 0$. Linearizing the equation about this state gives $(x = x^* + u)$

$$\ddot{u} + au = 0$$

where

$$a = \frac{d^2U}{dx^2}(x^*)$$

If x^* is a maximum (with $a < 0$), then the characteristic equation,

$$\lambda^2 + a = 0$$

has two real roots, one positive and one negative. If x^* is a minimum for U (with $a > 0$), then both roots are imaginary.

Solutions can be visualized by thinking of $U(x)$ as describing the profile of a surface on which a marble is rolling under the force of gravity. The marble's potential energy is proportional to $U(x)$, and its motion is described by Newton's law ($F = ma$). Obviously, if there is dissipation in the system (e.g., friction), then the marble will eventually roll to a valley floor, which is a (local) minimum of U. A marble starting exactly on a peak of U will remain there, but starting near a peak, it will roll away. However, the system studied here does not dissipate energy, and so a marble will not settle to a valley floor unless it starts there. It will either oscillate back and forth across a valley, be fixed at an extremum for U, or it will move to ∞.

Solutions can be plotted in the phase plane when we use x and $\dot{x}$ as variables. Since energy is conserved, we have

$$(dx/dt)^2/2 + U(x) = E$$

where E is the initial energy of the system. Figure 2.11 depicts two local minima of U and the corresponding phase portrait.

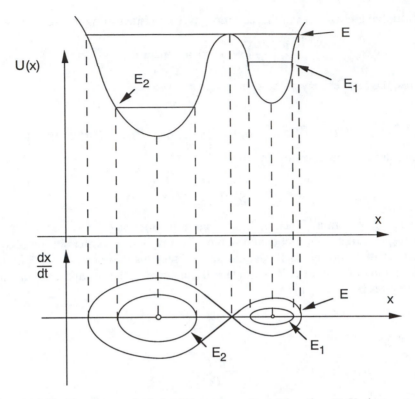

FIGURE 2.11. Phase portrait derived from the potential function U. Various energy levels are shown to correspond to various orbits in the phase plane.

Let $y = \dot{x}$, then the trajectory with energy E_2 lies on the graph of the function $y^2 = 2[E_2 - U(x)]$. This curve is defined for an interval $A \le x \le B$. y^2 measures the distance between the constant $U = E_2$ and the potential energy $U = U(x)$. Thus, the orbit is closed and so represents a periodic solution of the system.

Closed horizontal lines in the graph of U correspond to periodic solutions in the phase plane, and horizontal lines tangent at a local maximum of U correspond to separatrices. If such a line is bounded, the separatrix is *homoclinic*, that is, it goes from the saddle point and returns to it. If it is tangent to two maxima, the separatrix is *heteroclinic*, going from one saddle to the other.

Example. Consider the quartic potential

$$U(x) = \frac{x^4}{4} - a\frac{x^2}{2} + bx + c$$

for some constants a, b, and c. Figue 2.12 shows U and the corresponding phase portraits in two typical cases.

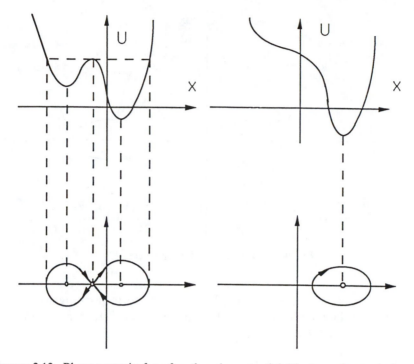

FIGURE 2.12. Phase portraits for a fourth-order potential. The two other typical cases are found by the change of variables $x \to -x$, which flips these plots through the y axis.

Example. Duffing's equation,

$$\ddot{x} + x - x^3 = 0$$

arises in many applications. For example, it is an approximation to a pendulum equation using

$$\sin y \sim y - y^3/3$$

and $x = y/\sqrt{3}$. In this case, $U(x) = x^2/2 - x^4/4$, so there are always three rest points, $x = 0$ and $x = \pm 1$. The first is a center and the second two are saddle points. It is interesting that the saddle–saddle connections can be found explicitly in this case. In particular, the function $x(t) = \tanh(t/\sqrt{2})$ is a solution of the equation that satisfies the boundary conditions $x(\pm\infty) = \pm 1$.

2.3.3   *Oscillation Period of $x'' + U_x(x) = 0$*

Consider the conservation equation

$$\frac{d^2 x}{dt^2} + \frac{\partial U}{\partial x}(x) = 0$$

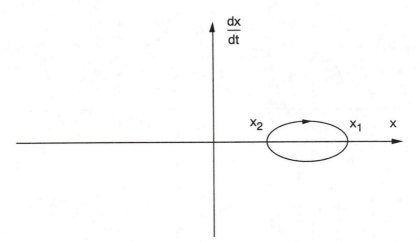

FIGURE 2.13. Closed orbit for $U(x) = E$.

Suppose that we set an energy level E and that corresponding to it is an oscillation as shown in Figure 2.13. The energy level E determines the values of x_1 and x_2, and solving

$$\frac{1}{2}\left(\frac{dx}{dt}\right)^2 + U(x) = E$$

on the top half of the orbit gives

$$\frac{dx}{dt} = \sqrt{2[E - U(x)]}$$

Integrating this over the top half of an orbit, as shown next, gives

$$\frac{1}{\sqrt{2}}\int_{x_1}^{x_2} \frac{dx}{\sqrt{E - U(x)}} = \frac{T}{2}$$

which is half the period. Thus, the period of the oscillation is

$$T = \sqrt{2}\int_{x_1}^{x_2} \frac{dx}{\sqrt{E - U(x)}}$$

2.3.4 Active Transmission Line

Figure 2.14 shows a series of cricuits that are coupled by resistors (R). In each circuit, there is a battery (voltage E) in series with a resistor (R^*). Each is in parallel with a capacitor (C) and a tunnel diode the current through which is given by $f(V)$.

The current coming into the node labeled V_i is I_{i-1}, and according to Kirchhoff's law [2.7], it is balanced by the total current going out of the

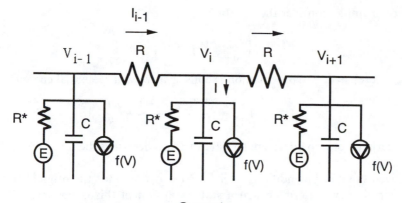

Ground

FIGURE 2.14. An active transmission line.

node: $I_i + I$. In turn, I splits into three components, one for each branch in the circuit:

$$I = f(V_i) + \frac{(V_i - E)}{R^*} + C\frac{dV_i}{dt}$$

On the other hand,

$$I_{i-1} = \frac{V_{i-1} - V_i}{R} \text{ and } I_i = \frac{V_i - V_{i+1}}{R}$$

Combining these formulas gives

$$C\frac{dV_i}{dt} = \frac{V_{i-1} - 2V_i + V_{i+1}}{R} - f(V_i) - \frac{V_i - E}{R^*}$$

A useful approximation views this as being nearly a continuum of circuits. We write $V(s, t)$ for the voltage at position s. Thus, $V_i(t) = V(i\delta s, t)$ where δs is the distance between circuit nodes. As the distance between nodes is decreased, the resistance R decreases. We suppose that $\delta s^2/R \sim D$ as $\delta s \to 0$. If $\delta s \ll 1$, then the model becomes (approximately)

$$C\frac{\partial V}{\partial t} = D\frac{\partial^2 V}{\partial s^2} - f(V) - \frac{V - E}{R}$$

If we look for static voltage profiles, we study the equation

$$0 = DV'' - f(V) - \frac{V - E}{R}$$

In the case of tunnel diodes, $f(V)$ is nearly a cubic curve, So this equation has the form of a conservation law with quartic potential.

For example, consider the equation

$$V'' + A\left(\frac{V^3}{3} - V\right) - \frac{V - E}{R} = 0$$

where here $' = d/ds$. This equation is conservative with potential energy

$$U(V) = A\frac{V^4}{12} - \frac{(1/R + A)V^2}{2} + \frac{EV}{R}$$

Our earlier work on quartic potential functions describes the solutions of this problem.

Given boundary conditions, such as $V(\pm\infty, t) = 0$ for the original transmission line, we determine that a static solution of this is possible only if $V = 0$ is a homoclinic saddle point. There are many static solutions that are periodic functions of s, but do not satisfy the boundary condition. The structure of solutions of the transmission line problem and their coordinated behavior as functions of t and s is rich, and it deserves further study [2.10].

The dynamics of solutions for $V(s, t)$ of the original model can also be studied using the phase plane. For example, if for each of a sequence of times the solution curve $V(s, t_n)$ is plotted on the V, V'' coordinates, then interesting aspects of the solution can be presented [2.11].

2.3.5   Phase–Amplitude (Angle–Action) Coordinates

It is often convenient to describe conservation models using the momentum $p = m\dot{x}$ rather than the velocity $\dot{x}$. At the same time, we change notation by replacing x by q. We define the Hamiltonian as

$$H(p, q) = p^2/2m + U(q)$$

The Euler-Lagrange equation can be now written in the form (see Section 2.1.1)

$$\frac{dq}{dt} = \frac{\partial H}{\partial p}$$

$$\frac{dp}{dt} = -\frac{\partial H}{\partial q}$$

This is a *Hamiltonian system* of equations corresponding to the physical problem whose energy is H [see 2.9].

Note that if $G(p, q)$ is a function that is invariant when evaluated along solutions of this system, then

$$0 = dG/dt = (\partial G/\partial p)\dot{p} + (\partial G/\partial q)\dot{q}$$

$$= (\partial G/\partial p)(\partial H/\partial q) - (\partial G/\partial q)(\partial H/\partial p)$$

$$\equiv [H, G]$$

$[H, G]$ is called the *Poisson bracket*, and so we see that any function that is invariant under the flow of solutions must commute with H in the sense that the Poisson bracket vanishes, and coversely.

This observation is a starting point for the development of Lie's algebraic theory for nonlinear oscillators. Lie's theory uses the product defined by the Poisson bracket to define an algebra. By viewing solutions as being actions in this algebra, we can accomplish replacing our original nonlinear problem by a linear one for representations of the Lie algebra [2.12].

Hamiltonian systems are *integrable*, that is, the structure of these equations often allows us to change variables in a way that produces solvable equations. This change of variables does not always exist, and, even when it does, it is not usually possible to carry it out. Still, there are two important cases where the work is straightforward, namely, the harmonic oscillator and the pendulum.

We first introduce a new amplitude variable a by the formula

$$a = H(p, q)$$

and we define a new phase variable ξ and a new frequency ω by solving the equations

$$\omega \frac{dq}{d\xi} = \frac{\partial H}{\partial p}(p, q)$$

$$\omega \frac{dp}{d\xi} = -\frac{\partial H}{\partial q}(p, q)$$

for $p = p(\xi, a)$, $q = q(\xi, a)$ subject to the constraints that $H(p, q) = a$. ω is chosen so that these solutions p and q have period 2π in ξ. This can be done near nondegenerate energy levels of the system, that is, near orbits that represent nontrivial periodic solutions, but this step is the first major hurdle in using the method.

The change of variables

$$p = p(\xi, a)$$

$$q = q(\xi, a)$$

is invertible since the Jacobian is

$$\det \begin{bmatrix} \dfrac{\partial p}{\partial \xi} & \dfrac{\partial p}{\partial a} \\ \dfrac{\partial q}{\partial \xi} & \dfrac{\partial q}{\partial a} \end{bmatrix} = -\frac{1}{\omega} \neq 0$$

Therefore, we can define inverse variables by

$$\xi = \xi(p, q)$$

$$a = H(p, q)$$

[2.7]. The second problem is finding these inverse variables in a useful form. When this can be done, it follows that a and ξ satisfy the equations

$$\frac{da}{dt} = 0$$

$$\frac{d\xi}{dt} = \omega(a)$$

which can be integrated directly. Because of this fact, Hamiltonian systems are referred to as being *integrable systems*. Since a is an amplitude and ξ is a phase variable, these are referred to as being *phase–amplitude*, or sometimes *angle–action*, variables.

Example. A pendulum. Chester's analysis of the pendulum [see 2.7] begins with the equation

$$\frac{d^2x}{dt^2} + \sin x = 0$$

This can be converted to a Hamiltonian system by setting q for x and

$$H(p, q) = \frac{p^2}{2} + 2\sin^2\frac{q}{2}$$

Then

$$dp/dt = -\sin q$$

$$dq/dt = p$$

We set $H(p, q) = 2a^2$, and we seek functions $p = p(\xi, a)$ and $q = q(\xi, a)$ and a constant $\omega(a)$, such that

$$\omega \, dq/d\xi = p$$

$$\omega \, dp/d\xi = -\sin q$$

where these functions are to have period 2π in ξ. We have

$$\left(\omega\frac{\partial q}{\partial \xi}\right)^2 = p^2 = 4\left(a^2 - \sin^2\frac{q}{2}\right)$$

so we study the two cases, $a < 1$ and $a > 1$, separately. In the first case, the pendulum oscillates regularly, and, in the second, it executes full cycles.

If $a < 1$, we set $\sin(q/2) = as$, and we get an equivalent integral equation for s

$$\omega \int_0^s (1 - \sigma^2)^{-1/2}(1 - a^2\sigma^2)^{-1/2} \, d\sigma = \xi$$

if $q = 0$ corresponds to $\xi = 0$. The inverse relation is

$$s = \frac{\sin(q/2)}{a} = \text{sn}\left(\frac{\xi}{\omega}, a^2\right)$$

where sn denotes the elliptic sine function [see 2.13].
This function is known to have period $4K(a^2)$ in ξ/ω where

$$K(a^2) = \int_0^1 \frac{ds}{\sqrt{(1-s^2)(1-a^2s^2)}} = \int_0^{\pi/2} \frac{d\phi}{\sqrt{1-a^2\sin^2\phi}}$$

In order that the function $\text{sn}(\xi/\omega, a^2)$ have period 2π in ξ, we make the choice

$$\omega = \pi/2K(a^2)$$

Note that as $a \to 0$, $K(a^2) \to \pi/2$ so the period of small amplitude oscillations is 2π.
For $a > 1$, we define $\sin(q/2) = s$, so

$$\omega \int_0^s \frac{d\sigma}{\sqrt{(1-\sigma^2)[1-(\sigma/a)^2]}} = a\xi$$

and

$$s = \sin\frac{q}{2} = \text{sn}\left(\frac{a\xi}{\omega}, a^{-2}\right).$$

This function has period $4K(a^{-2})$ in $a\xi/\omega$, or 2π in ξ if

$$\omega = \pi a/2K(a^{-2})$$

In this case, the pendulum is executing full oscillations. In either case, we have derived $\omega(a)$ for the integrable form of the pendulum equations, as well as the change of variables that takes the pendulum to integrable form. Explicit formulas for the oscillations can be found by solving for q in each case.

2.3.6 *Conservative Systems with N Degrees of Freedom*

A system of N particles will have N position and N momentum coordinates. Say particle i has position $q_i(t)$ at time t and (generalized) momentum $p_i(t)$. If the potential energy is

$$U(q_1, \cdots, q_N)$$

then the Hamiltonian is given by the formula

$$H(p, q) = \sum_{i=1}^N \frac{p_i^2}{2m_i} + U(q)$$

and the equations of motion are

$$dq_i/dt = \partial H/\partial p_i$$
$$dp_i/dt = -\partial H/\partial q_i$$

for $i = 1, \cdots, N$. Thus, $2N$ equations result which are equivalent to the N second-order equations

$$\ddot{q}_j + \frac{\partial U}{\partial q_j}(q) = 0$$

for $j = 1, \cdots, N$.

This is a starting point for many studies in statistical and celestial mechanics [2.14, 2.15]. In addition, these equations arise as characteristic equations in solving first-order nonlinear partial differential equations. They also arise in the calculus of variations where they appear as Euler-Lagrange equations. Finally, an extensive geometrical theory, a symplectic geometry that is based on differential invariants of this system, has evolved. Because of the central role played by such systems in a wide variety of mathematical disciplines, these equations have been and continue to be extensively studied. We study examples of this general system further in later chapters. Here we note the following useful connection between Hamiltonian systems and certain partial differential equations [see 2.9].

2.3.7 Hamilton-Jacobi Theory

Hamiltonian systems are closely related to certain first-order partial differential equations. On one hand, let $\phi(t, x, a)$, where $x \in E^N$, be a smooth solution of the Hamilton-Jacobi equation

$$\partial \phi / \partial t + H(x, \nabla \phi) = 0$$

for $t \geq 0$. We suppose that ϕ contains N parameters $a_1, \cdots, a_N$ such that

$$\det \frac{\partial^2 \phi}{\partial x_i \partial a_j} \neq 0$$

Then ϕ is called a *complete integral* of the equation. We suppose here that $H(x, p)$ is a continuously differentiable function of the $2N$ variables $x_1, \cdots, x_N$, $p_1, \cdots, p_N$. Then the equations

$$\partial \phi / \partial a_j = b_j$$

where $b_1, \cdots, b_n$, are free constants, can be used to determine x as a unique function of t, a, and b.

Let us define

$$p_j = \partial \phi / \partial x_j$$

then these functions satisfy the ordinary differential equations

$$dx_j / dt = \partial H / \partial p_j$$

$$dp_j / dt = -\partial H / \partial x_j$$

for $j = 1, \cdots, N$ [see 2.16]. This shows how solutions of a Hamilton-Jacobi partial differential equation can define solutions of an associated Hamiltonian system.

On the other hand, a Hamilton-Jacobi equation

$$\partial\phi/\partial t + H(x, \nabla\phi) = 0$$

with initial conditions

$$\phi(0, x) = g(x) \qquad \text{and} \qquad \nabla\phi(0, x) = \nabla g(x)$$

for a given smooth function $g(x)$ can be solved in the following way. Define variables

$$p_j = \partial\phi/\partial x_j \qquad \text{for} \qquad j = 1, \cdots, N,$$

and for each $\xi \in E^N$, let $x(t, \xi)$, $p(t, \xi)$ denote solutions of the Hamiltonian system that satisfy the initial conditions

$$x_j(0) = \xi_j \qquad p_j(0) = \frac{\partial g}{\partial x_j}(\xi_j)$$

Finally, let $\Phi(t, \xi)$ denote the solution of the equation

$$\frac{d\Phi}{dt} = p(t, \xi) \cdot \frac{\partial H}{\partial p}[x(t, \xi), p(t, \xi)] - H$$

with initial condition $\Phi(0, \xi) = g(\xi)$. The change of variables $x = x(t, \xi)$ can be inverted: Say, $\xi = \mathring{\xi}(t, x)$. Then we define

$$\phi(t, x) = \Phi[t, \mathring{\xi}(t, x)]$$

This function satisfies the Hamilton-Jacobi equation

$$\partial\phi/\partial t + H(x, \nabla\phi) = 0$$

and initial conditions

$$\phi(0, x) = g(x), \qquad \nabla\phi(0, x) = \nabla g(x)$$

We say that the Hamiltonian system forms the set of *characteristic equations* for this Hamilton-Jacobi equation.

These two calculations show the relation between a Hamiltonian system and its associated Hamilton-Jacobi equation. There are other interesting connections between these systems with interpretations in the calculus of variations [see 2.9]. The following example illustrates Jacobi's method.

Example. A two body problem. If bodies of masses m_1 and m_2 are placed at points x and $y \in E^3$, respectively, then Newton's Law of Gravitation states that the potential energy of the pair is

$$U(x, y) = am_1 m_2/r$$

where $r = |x - y|$ and a is a constant. The equations of motion of this system are

$$m_1\ddot{x} = -\nabla_x U$$

$$m_2\ddot{y} = -\nabla_y U$$

The motion can be shown to lie in a plane, and for convenience, we place the second mass at the origin of our coordinate system. The result is a system of two equations for two of the components of $\mathbf{x}$, say x_1 and x_2:

$$m_1\ddot{x}_1 + \frac{\partial U}{\partial x_1} = 0$$

$$m_2\ddot{x}_2 + \frac{\partial U}{\partial x_2} = 0$$

where now

$$U(x_1, x_2) = \frac{am_1 m_2}{\sqrt{x_1^2 + x_2^2}}$$

This is equivalent to a Hamiltonian system. Let

$$H(x_1, x_2, p_1, p_2) = \frac{p_1^2 + p_2^2}{2} - U(x_1, x_2)$$

The corresponding Hamilton-Jacobi equation is

$$\frac{\partial \phi}{\partial t} + \frac{1}{2}\left(\frac{\partial \phi}{\partial x_1}\right)^2 + \frac{1}{2}\left(\frac{\partial \phi}{\partial x_2}\right)^2 = \frac{b}{\sqrt{x_1^2 + x_2^2}}$$

where $b = am_1 m_2$. In polar coordinates, this equation becomes

$$\frac{\partial \psi}{\partial t} + \frac{1}{2}\left(\frac{\partial \psi}{\partial r}\right)^2 + \frac{1}{2r^2}\left(\frac{\partial \psi}{\partial \theta}\right)^2 = \frac{b}{r}$$

where $\psi(t, r, \theta) = \phi(t, r\cos\theta, r\sin\theta)$. This equation can be solved as a sum of functions, one in each variable. The result is that

$$\phi = a_1 t + a_2 \theta + g(r, a_1, a_2)$$

where

$$g = \int_{r_0}^{r} \sqrt{2br^{-1} - 2a_1 - a_2^2 r^{-2}}\, dr$$

Solving the equations

$$b_1 = \partial\phi/\partial a_1, \qquad b_2 = \partial\phi/\partial a_2$$

leads to the formula

$$r = c[1 - e^2 \sin(\theta - b_2)]^{-1}.$$

which defines a solution of the two body problem.

2.3.8 *Liouville's Theorem*

Consider the system of differential equations

$$dx/dt = f(x), \qquad x(0) = \xi,$$

where x, ξ, and f are in E^N and f is a smooth function of the components of x. Let us consider the ensemble of solutions beginning in a volume $V(0)$. The size of this set after t time units will be given by the integral

$$\int_{V(t)} dx$$

where $V(t)$ is a collection of points

$$\{x = x(t, \xi) \colon \xi \in V(0)\}$$

If we make the change of variables $\xi \to x$ by setting $x = x(t, \xi)$, then we have

$$\int_{V(t)} dx = \int_{V(0)} \det\left(\frac{\partial x}{\partial \xi}\right) d\xi$$

$\partial x/\partial \xi$ can be determined from the original equation by solving the problem

$$\frac{d}{dt}\frac{\partial x}{\partial \xi} = \frac{\partial f}{\partial x}(x)\frac{\partial x}{\partial \xi}, \qquad \frac{\partial x}{\partial \xi}(0) = \text{identity}$$

Thus, the function $y(t) = \det(\partial x/\partial \xi)(t)$ satisfies

$$\dot{y} = \text{tr}[f_x(x)] y, \qquad y(0) = 1$$

The coefficient of the right-hand side of this equation is the divergence of f, $\text{div } f = \nabla \cdot f$. It follows that

$$\int_{V(t)} dx = \int_{V(0)} \exp\left\{\int_0^t \text{div } f[x(t', \xi)] \, dt'\right\} d\xi$$

If $\text{div } f = 0$, then volume $V(0) = $ volume $V(t)$ for any $t > 0$, since these measures are defined by the same integral. With these calculations, we have proved the following useful theorem.

   Liouville's Theorem. *Let the conditions listed above on f be satisfied. Then the flow defined by the differential equaton $dx/dt = f(x)$ is volume preserving if $\text{div } f(x) = 0$ for all x.*

   This result shows that Hamiltonian systems define volume-preserving flows. In fact, the divergence of the right-hand side of a Hamiltonian system is zero, since

$$\sum_{i=1}^{N} \frac{\partial q_1'}{\partial q_i} + \sum_{i=1}^{N} \frac{\partial p_1'}{\partial p_i} = \sum_{i=1}^{N} \frac{\partial^2 H}{\partial p_i \partial q_i} - \sum_{i=1}^{N} \frac{\partial^2 H}{\partial p_i \partial q_i} = 0$$

Therefore, for each t the solutions of a Hamiltonian system define a mapping of E^N into itself for which the Lebesgue measure is invariant.

2.4   Dissipative Systems

We have seen that conservative systems typically have a rich structure of oscillatory solutions. However, systems that do not conserve energy are different. For example, oscillations of such systems are usually isolated, as for Lienard's equation in Section 2.1.3. Two important examples are described in detail in this section: van der Pol's equation and the VCON model.

2.4.1   *Van der Pol's Equation*

Recall that van der Pol's equation is

$$\ddot{x} + A(x^2 - 1)\dot{x} + x = 0$$

We know from Section 2.1.3 that, for any choice of $A > 0$, there is a unique periodic solution of this equation, and all other solutions approach it as $t \to \infty$ except for the unstable equilibrium at $x = 0$, $\dot{x} = 0$.

It is difficult to estimate the period of this oscillation. After some work on perturbation methods, we see that the period for large values of A is comparable to A and for small values of A the period is comparable to 2π. Figure 2.15 shows the oscillation for two values of A.

2.4.2   *VCON*

The voltage controlled oscillator neuron model (VCON) is a model nerve cell that is similar to the phase-locked loop of Section 2.2.4:

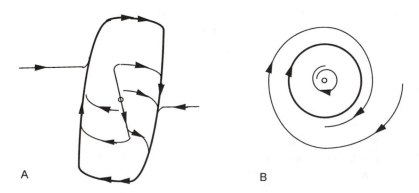

A   B

FIGURE 2.15. (a) $A = 5.0$. $x = x' = 0$ is an unstable node. The double arrows indicate branches that are traversed rapidly. (b) $A = 0.1$. The limit cycle is approximately a circle and its period is nearly 2π.

$$dx/dt = \omega + z$$

$$\tau \, dz/dt = -z + W \cos_+ x$$

In this section, we study the free case of this system when $W = A$ is a positive constant.

With $y = \omega + z$, the system takes the form

$$dx/dt = y$$

$$dy/dt = -\sigma y + I + f(x)$$

where $\sigma = 1/\tau$ and $I = \omega/\tau$ are constants and $f(x) = (A/\tau) \cos_+ x$.

This problem can be completely analyzed. First, all solutions approach the strip

$$S = \{(x, y) : I - f^* \le y \le I + f^*\}$$

where $f^* = \max_x |f(x)| = A/\tau$. In fact, we see directly from the second equation that solutions approach this strip at an exponential rate with time constant σ. It is appropriate to view this system as being on the cylinder $|y| < \infty$, $0 \le x \le 2\pi$, since the equations are invariant under the translation $x \rightarrow x + 2\pi$. We consider several cases.

$|I| > f^*$

When $I > f^*$, there are no static states since the isoclines $dx/dt = 0$ and $dy/dt = 0$ never cross. The fact that the strip S is invariant shows that the mapping defined from the cross-section of this strip at $x = 0$ to the one at $x = 2\pi$ has a fixed point in S, and so there is a periodic solution, say $y = Y(x)$. This periodic solution is a limit cycle on the cylinder. We prove this in the following theorem [2.8].

Theorem. If $I > f^*$, there is a unique limit cycle on the cylinder and all solutions approach it as $t \rightarrow \infty$.

Proof. First, we need only consider the strip S. Since S is invariant, the solutions define a mapping of the cross-section $x = 0$ of S into the one for $x = 2\pi$. This mapping is also a contraction mapping, which follows from the following short calculation showing that the mapping has a positive derivative that is less than one.

Let $M(y_0)$ denote the value of y when $x = 2\pi$ given that $y = y_0$ when $x = 0$. We can write y as a function of x since there are no critical points in S, and we have

$$\frac{dy}{dx} = \frac{-\sigma y + I + f(x)}{y}$$

$y(0) = y_0$. The derivative $\partial y / \partial y_0$ satisfies the linear problem

$$\frac{d}{dx}\frac{\partial y}{\partial y_0} = -\left[\frac{I + f(x)}{y^2}\right]\frac{\partial y}{\partial y_0}$$

$$\frac{\partial y}{\partial y_0}(0) = 1$$

Therefore,

$$M'(y_0) = \exp\left[-\int_0^{2\pi}\frac{I + f(x)}{y^2}dx\right] < 1$$

It follows that this transformation has a unique fixed point, and it lies on a limit cycle, say L. The stability of this limit cycle follows from the stability of S and the fact that L is the largest invariant sets in S.   ■

If we denote the limit cycle L by $y = Y(x)$, then the relation between x and t on this limit cycle can be determined by integrating

$$dx/Y(x) = dt$$

In fact, the period is

$$T = \int_0^{2\pi}\frac{dx}{Y(x)}$$

$|I| < f^*$

In this case, there are two equilibria on the cylinder, namely, when $f(x) = -I$. The left one is denoted by x_L and for it $f'(x_L) < 0$. The other is denoted by x_R, and $f'(x_R) > 0$. The linearization of the system about these states has the form

$$\frac{d^2x}{dt^2} + \sigma\frac{dx}{dt} - f'(x^*)x = 0$$

The characteristic polynomial is $\lambda^2 + \sigma\lambda - f'(x^*) = 0$, and its roots are

$$\lambda = -\frac{\sigma}{2} \pm \sqrt{\frac{\sigma^2}{4} + f'(x^*)}$$

Thus, we see that $x^* = x_L$ is a stable sink. It is a stable spiral point if I is small, and it is a stable node if I is near f^*, since $f'(x_L) \sim 0$. Similarly, we see that x_R is a saddle point. The following Theorem describes stability properties of the system when $|I| < f^*$.

Theorem. *For each value of τ and each $I < f^*$, there is a number $\sigma^* = \sigma^*(\tau, I)$ such that*

1. *If $\sigma < \sigma^*$, then the sink is globally stable on the cylinder, with the exception of separatrices.*
2. *If $\sigma^* < \sigma < f^*$, a stable sink and a stable limit cycle both exist.*

The proof of this Theorem is given in Ref. 2.8 for $f(x) = \cos x$. It is accomplished by showing that, if σ is small, then the stable and unstable manifolds are as in Figure 2.16, and for σ large as in Figure 2.17. σ^* is the value at which there is a saddle–saddle connection, that is, a heteroclinic point.

This system exhibits *hysteresis* as ω increases from zero to 1.0 and back to zero. For small values of ω, the system equilibrates to a stable node, where it

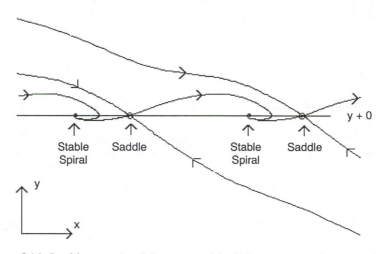

FIGURE 2.16. In this case, $I = 0.5$, $\tau = \sigma = 1.0$, $f(x) = \cos_+ x$, and we see that all solutions approach stable nodes except for the stable manifolds of the saddle points. Thus, the VCON output equilibrates to a stable voltage, $V(x_L)$.

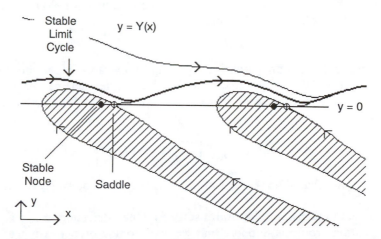

FIGURE 2.17. In this case, $I = 0.75$, and the other data are as in Figure 2.16. We see that there are stable nodes and a stable periodic solution, $y = Y(x)$, for this choice of parameters. Y is referred to as being a *running periodic solution* or a limit cycle on the cylinder. The running periodic solution corresponds to repetitive firing by the cell, and the VCON has two stable modes of behavior in this case.

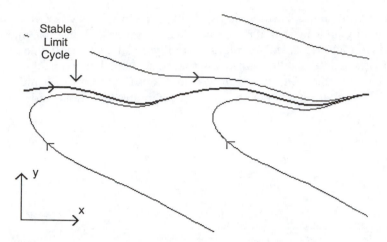

FIGURE 2.18. In this case, $I = 1.0$, and the other data are as in Figure 2.16. There are no equilibria, but there is a unique limit cycle, and it is globally stable.

remains until the saddles and nodes coalesce and disappear (at $\omega = f^*$). After this, the limit cycle is reached as shown in Figure 2.18. When ω is decreased, the system remains on the limit cycle until the saddle–saddle connection is reached. Below this, the solution equilibrates to a sink.

Gradient Systems

Let $F(\mathbf{x})$ denote a real valued mapping from $\mathbf{x} \in E^N$ to E^1. The gradient of F, $\nabla F(\mathbf{x})$, defines a vector that is normal to the level surface $F(\mathbf{x}) = c$ at the point x. Thus, solutions of the equation

$$d\mathbf{x}/dt = \nabla F(\mathbf{x})$$

are curves in E^N that are parallel to these normal vectors. Because of this, F plays the role of an energylike function. For example, there can be no periodic solution to such a system, since a solution having a least period $T > 0$ will satisfy

$$0 = F[x(T)] - F[x(0)] = \int_0^T \frac{dx}{dt} \nabla F(\mathbf{x}) \, dt = \int_0^T |\nabla F(\mathbf{x})|^2 \, dt$$

which implies that $\nabla F(\mathbf{x}) = 0$ for all t. Thus, $x(t)$ must be an equilibrium of the system.

Gradient systems arise in several settings. One interesting source is from optimization problems. Suppose that we want to find proper extrema of the function $F(\mathbf{x})$. We find these among the values of x for which $\nabla F(\mathbf{x}) = 0$. Thus, we solve either of the associated gradient problems

$$\dot{\mathbf{x}} = \pm \nabla F(\mathbf{x})$$

using an initial guess $x(0)$. If this solution approaches a static state, the limit will be a candidate for an extremum. We do this for many initial points, and build up a library of possible extrema that must be tested separately [2.17].

2.4.3 Cusp Catastrophe

The cusp catastrophe surface provides a number of interesting examples of nonlinear oscillations. Consider the quartic polynomial

$$F(x) = \frac{x^4}{4} - a\frac{x^2}{2} + bx + c$$

where a, b, and c are constants. Note that this form includes all quartics since the coefficient of x^3 is the sum of the roots and by a translation this can be made zero. We study the solutions of the equation

$$\dot{x} = -\partial F/\partial x$$

First we find the critical values, that is, the values of x for which $\partial F/\partial x = 0$. To do this, we must solve the equation

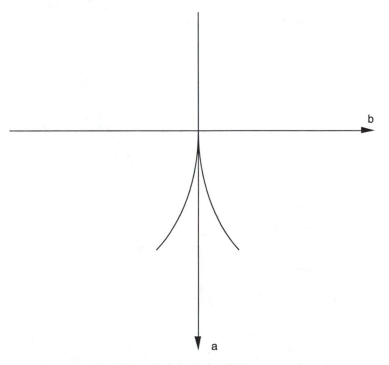

FIGURE 2.19. Bifurcation diagram of the cusp catastrophe.

$$x^3 - ax + b = 0$$

in a way that describes all solutions for any choices of a and b.

There are either one or three real roots, and they can be found by solving for b and plotting the resulting surface:

$$b = ax - x^3.$$

Extreme values of this function of a and x occur where $\partial b/\partial x = 0$, that is, $a - 3x^2 = 0$. Thus,

$$b = \pm \frac{2a^{2/3}}{3\sqrt{3}}$$

The result is depicted in Figure 2.19.

On the other hand, the ax trace of the solution when $b = 0$ is a parabola $a = x^2$ and the line $x = 0$. Combining these facts leads to the surface plotted in Figure 2.20. This is called the *cusp catastrophe surface* [see 2.18].

The upper and lower branches of this surface are stable equilibria for the system

$$\dot{x} = -\partial F/\partial x$$

Interesting oscillators can be generated by allowing a and b to change with time, and the following examples illustrate two of them.

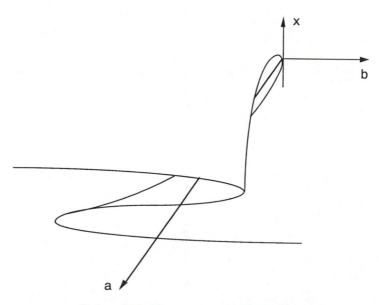

FIGURE 2.20. The cusp catastrophe surface.

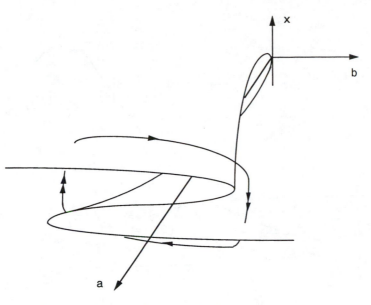

FIGURE 2.21. Van der Pol oscillation on a cusp surface.

Van der Pol's Equation

Consider the system

$$\dot{x} = -\partial F/\partial x$$
$$\dot{a} = 0, \qquad a(0) > 0$$
$$\dot{b} = \varepsilon x$$

where ε is a small parameter, $\varepsilon \ll 1$. Figure 2.21 shows the solutions of this system. Since $\varepsilon \ll 1$, the solutions equilibrate to the cusp catastrophe surface faster than b changes. Therefore, the solutions appear to be glued to this surface except when they cross an edge.

Lock-Washer Oscillator

Consider the system

$$\dot{x} = -\partial F/\partial x$$
$$\dot{a} = -\varepsilon b$$
$$\dot{b} = \varepsilon a$$

Now, the point $[a(t), b(t)]$ executes a slow circle in the ab plane, and the x variable is attracted to the stable branch of the cusp catastrophe surface lying

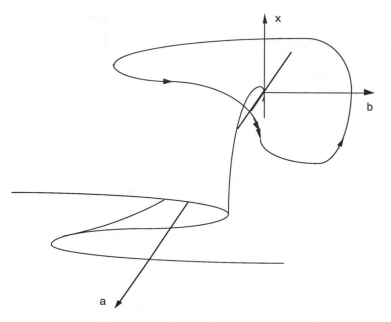

FIGURE 2.22. Lock-washer oscillation on a cusp surface.

above or below the circle. Figure 2.22 shows a typical solution of this problem.

Many other oscillations can be visualized using this surface by adjusting
the dynamics of a and b. This surface occurs in our analysis of various aspects
of nonlinear oscillations.

2.5   Stroboscopic Methods

If we trace an oscillation using a flashing light, then the sequence of points
observed at the flashes defines a table of values for an iteration of E^N into
itself. Little is known about such iterations when $N \geq 3$.

So, we restrict attention here to examples of interval mappings, circle mappings, and mappings of the plane. Most of these topics are of independent
interest, but we restrict our attention in most cases to problems related directly to oscillations.

2.5.1   *Chaotic Interval Mappings*

Iterations of a real-valued function can be surprisingly complicated. Even
simple functions can have highly irregular iterates, and similar behavior is
found in solutions of certain nonlinear oscillator problems. We begin with

definitions and a summary of some relevant work on function iteration. This is followed by some computer experiments for iterations.

Consider a continuous function g that maps a finite interval $I = (a, b)$ into itself. We define iterates of g by the notations $g^0(x) = x$, $g^1(x) = g(x)$, and $g^m(x) = g[g^{m-1}(x)]$ for $m = 2, 3, 4, \cdots$. We say that x^* is a *fixed point of period m* if $g^m(x^*) = x^*$, but $g^j(x^*)$ is not equal to x^* for $j < m$. The *orbit* of a point y is the set of points

$$\Omega(y) = \{g^j(y): j = 0, 1, 2, \cdots\}$$

Finally, we say that x^* is a *stable fixed point* of g if (1) $g(x^*) = x^*$, and (2) for every point y near x^* we have that

$$\lim_{n \to \infty} g^n(y) = x^*$$

Sarkovski's Sequence

A striking result found by Sarkovski describes interesting things about fixed points of g [2.19]. *Sarkovski's sequence* is defined by rearranging the natural numbers as shown next

$$1, 2, 2^2, \cdots, 2^n, \cdots \# \cdots, 9 \cdot 2^n, 7 \cdot 2^n, 5 \cdot 2^n, 3 \cdot 2^n, \cdots, 9 \cdot 2, 7 \cdot 2, 5 \cdot 2, 3 \cdot 2, \cdots,$$
$$9, 7, 5, 3$$

The symbol $\#$ separates powers of 2 (pure harmonics) coming from the left from partials (odd numbers and their harmonics) coming from the right.

Sarkovski showed that if g has a fixed point of order p, then it has a fixed point of each order q lying to the left of p in this sequence. Thus, if there is a fixed point of period two, then there is also a fixed point; at the other extreme, the presence of a fixed point of period 3 implies that there are fixed points of all periods [see also 2.20].

In one sense, Sarkovski's sequence describes the *complexity* of invariant sets for g. The farther along g is in the sequence, the richer the collection of sets is that are invariant under it.

An interesting approach to iteration of smooth functions is described by Ulam [2.21]. Under certain circumstances there is a function (called a *density function*) that is invariant under g. A simple example of this is when

$$g = \min(2x, 2 - 2x)$$

and $I = [0, 1]$. In this case, the inverse image of an interval of length L is the union of two intervals of total length L. Therefore, the density function $d = 1$ is invariant. We say in this case that Lebesgue's measure is invariant under g. Invariant measures describe for interval mappings the idea of volume preserving for Hamiltonian flows [see Section 2.3.8].

The ergodic theorem [2.22] has a useful implication for this simple example. It states (roughly) that if Lebesgue's measure is invariant, iterates under g of most initial points move all over the interval. In particular, the probability of an iterate hitting a given subinterval is proportional to the length of that subinterval. This is the basis of the work in Ref. 2.23, which is described next.

Iteration Histogram and Entropy

The ergodic theorem suggests a method for using a computer to experiment with function iterates. In particular, let us partition the invariant interval I into M equal parts (cells) and select a point from I at random. This point is iterated N^* times under g and the number of the iterates that hit each cell is recorded. If N^* and M are sufficiently large, then in cases where the ergodic theorem applies (in particular, when there is an invariant measure that is absolutely continuous with respect to Lebesgue's measure) the histogram that results from the iteration experiment reflects the graph of an invariant density function.

Suppose that the iteration histogram has cell contents $c_1, \cdots, c_M$. If we normalize these by setting

$$p_j = c_j \bigg/ \sum_{k=1}^{M} c_k$$

then $(p_1, \cdots, p_M)$ defines a probability distribution. The number

$$H(p) = - \sum_{j=1}^{M} p_j \log p_j$$

is called the *entropy* of the distribution described by p.

In the above equation, H indicates how spread out are the components of p. For example, if the components of p were approximately normally distributed, we would have $p_j \sim \exp[-(j-m)^2]$, and for this H would be a second moment of the distribution. Therefore, it is related to p's variance. If all cells are equally filled ($p_j = 1/M$), then $H(p) = \log M$, which is the maximum value of H. If only one cell is filled ($p_J = 1$), then $H(p) = 0$, which is the minimum value of H.

Rather than plotting a density function's histogram as in Ref. 2.23, we will simply indicate which cells are eventually hit by the iteration. Figure 2.23 shows the result of the iteration experiment performed for the function

$$g(x) = rx(1 - x).$$

In Figure 2.23, the value $g(1/2)$ is iterated 2,000 times and the first 100 iterates are discarded to suppress transients. The value $x = 1/2$ is chosen for the initial point of the iteration because it is known that for any choice of r the orbit of this point evolves into the invariant set of g that attracts most initial points. We plot also the entropy of the iteration's distribution to quantify its complexity.

In Figure 2.23, the unit interval, $0 \le x \le 1$, is partitioned into 300 subintervals. The dark pixels (picture elements) are those subintervals in

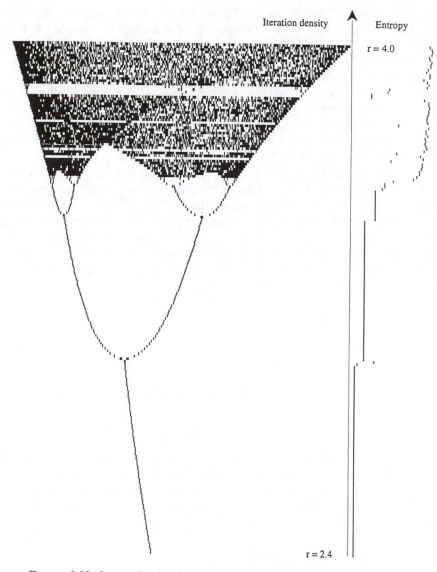

Iteration density   Entropy

r = 4.0

r = 2.4

FIGURE 2.23. Iterate density and entropy as computed for $g(x) = rx(1 - x)$.

which some iterates of the mapping lie. Figure 2.23 illustrates high-resolution computation followed by pixel truncation imposed by the plotting device. In this, we choose 300 values of r between $r = 1$ and $r = 4$, and for each we plot horizontally the pixels containing at least one iterate of g. We could successively rescale the viewing window and get successively more refined views of the iteration since the computations are of greater accuracy than the plotting device used here. In Figure 2.23, all cells having at least one iterate are plotted for each of 300 values of r between 1 and 4. We see first ($1 \leq r \leq 3$) the stable fixed point $x = (r - 1)/r$. Beyond $r = 3$, there is a cascade of new periodic orbits appearing that correspond to Sarkovski's sequence lying to the left of #. The # occurs approximately at $r \approx 3.57$ for this function. The left and right branches of this diagram come together at $r = r^* \approx 3.68$. The behavior beyond # ($r > 3.57$) is often referred to as being *chaotic*, although r intervals where there are stable 6, 5, and 3-period orbits are clearly visible even at the level of resolution of this experiment. The term *chaotic* is not cleary defined, and it is used with different meanings in various contexts. We take it to mean highly irregular behavior.

The experiment described in Figure 2.23 shows that the simulation is useful when there is a single fixed point of g, a case to which the ergodic theorem does not apply directly. In such a case, there is an invariant measure whose density function is a Dirac delta function. Still, the iteration experiment produces a quantized approximation to the Dirac function since the iterates all accumulate in the cell containing the fixed point. This observation enables us to uncover Sarkovski's sequence from function iteration, up to the resolution of the output device, as shown in Figure 2.23.

Finite Semigroup Approximation

Some insight to the complicated dynamics described above can be found by approximating the function g by a pixel approximation. Therefore, we partition the interval $[0, 1]$ into N equal subintervals, $J_1 = [x_0, x_1]$, $J_2 = (x_1 x_2]$, $\cdots$, $J_N = (x_{N-1}, x_N]$. We define an integer-valued function f by the table of values

$$f(i) = j \quad \text{if } g(x_{i-1}) \in J_j \quad \text{for } i, j = 1, \cdots, N - 1.$$

Thus, $f(i) = j$, if g maps the left end point of J_i into J_j and f defines a mapping of the set $S = \{1, \cdots, N\}$ into itself. It also describes the graph of g when N pixels are used to plot it. The iterates of f define a finite semigroup [2.24]. Figure 2.24 shows the result of performing the iteration experiment just described for f rather than g.

Markov Chain Approximation to g

What has this to do with random behavior of iterates of g? Using f, we can associate with g a Markov chain. Let us define the graph matrix $G[f]$ by the formulas

$$G_{j,k} = 1 \quad \text{if } f(j) = k$$

$$G_{j,k} = 0 \quad \text{otherwise}$$

for $j, k = 1, \cdots, N$. The matrix G is a stochastic matrix since each of its row sums is equal to one and each of its components is nonnegative. The graph

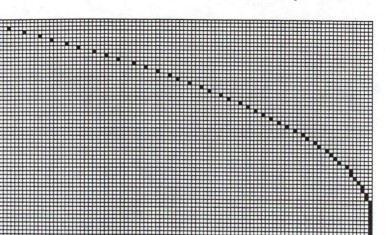

(a)

FIGURE 2.24a. Graph transition matrix for $g(x) = 3.98\,x(1 - x)$ using 100 pixels.

of f appears in $G[f]$ where it is traced by the ones appearing in that matrix as shown in Figure 2.24.

The spectrum of G consists of 1s, 0s, and some roots of unity. These eigenvalues are closely related to the dynamics of f. In fact, each nontrivial root of unity eigenvalue corresponds to a periodic orbit of f, and the corresponding eigenvector consists of 1s lying over a periodic orbit of f. Similarly, each unit eigenvalue (those not among roots-of-unity eigenvalues) corresponds to a fixed point of f and each zero eigenvalue corresponds to a transient state of f.

Finally, we can associate with g and N another Markov chain. We define M_{jk} to be the proportion of J_j that maps into J_k under g. This Markov chain presumably better approximates g than does f or $G[f]$. Figure 2.26 shows the results of a Monte Carlo simulation based on this Markov chain.

The complexity of these iterations can be described using the histogram entropy. For example, let the eigenpairs for $G[f]$ be denoted by λ_k, ϕ_k for $k = 1, \cdots, N$. Add up all of the eigenvectors ϕ_k for which $\lambda_k \neq 0$

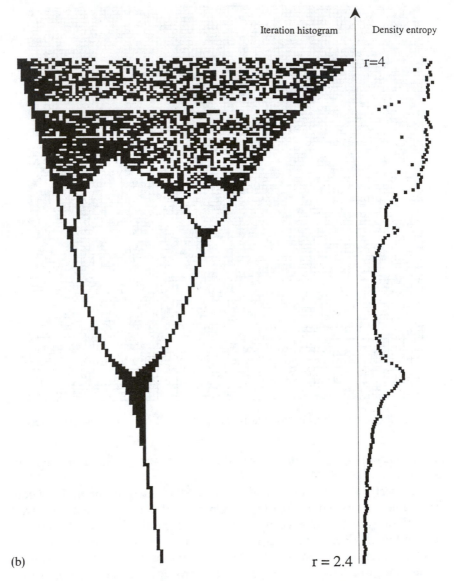

FIGURE 2.24b. Iteration of the pixel approximation to g using 100 pixels.

$$\phi = \sum_{\lambda_k \neq 0} \phi_k$$

Normalize this vector by its length, say $\mathbf{q} = \phi/|\phi|$, so $\mathbf{q} = (q_1, \cdots, q_N)$ is a probability vector.

The entropy $H(\mathbf{q})$ describes the complexity of the iterates of f, and so gives an indication of the dynamics of g. A similar statistic can be computed for the Markov chain M. (see [7.7]).

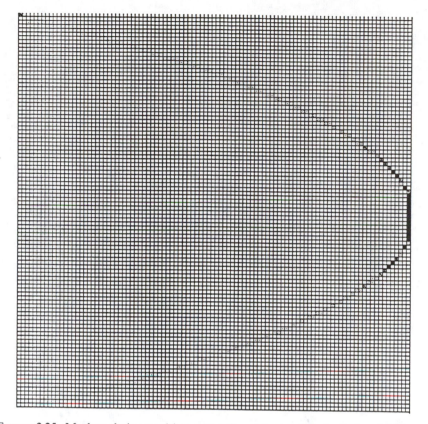

FIGURE 2.25. Markov chain transition matrix for $g(x)$ using 100 pixels. ($r = 3.98$) Row sums add to 1.0, and shading indicates the relative density of each component.

2.5.2 Circle Mappings

Consider now a system of differential equations on a torus

$$\dot{x} = F(x, y)$$

$$\dot{y} = G(x, y)$$

where F and G are differentiable functions that are doubly periodic, say

$$F(x + 2\pi, y) = F(x, y) = F(x, y + 2\pi)$$

for all x and y. This is a system of equations on a torus, as discussed in Section 2.2.2. If $G \neq 0$, then we can divide the first equation by the second and replace y by t:

$$\dot{x} = f(t, x)$$

where now $f(t, x) = F(t, x)/G(t, x)$ is a 2π-periodic function of t and x. Thus, certain systems on the torus are equivalent to periodically forced scalar problems.

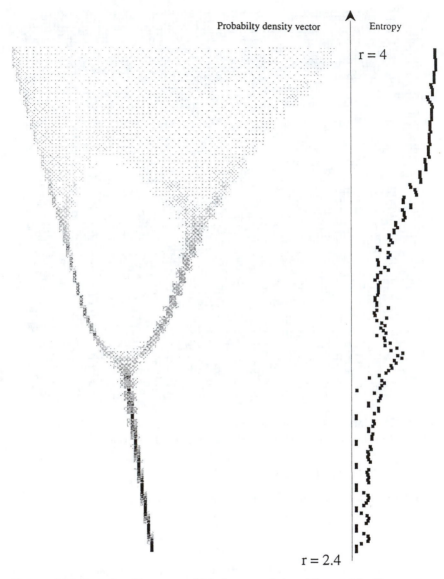

FIGURE 2.26. Iteration histogram of Markov graph transition matrix. Row sums on left figure add to 1.0, and gray scale indicates the relative density in each.

Poincare's mapping of solutions is defined by

$$\phi: x(0) \to x(2\pi) \equiv \phi[x(0)]$$

and it corresponds to a mapping of the circle $y = 0$ on the torus into itself.
Consider a function $Y(t)$ for which

$$\frac{dY}{dt} = f(t, Y) \qquad \text{and} \qquad Y(0) = x(0) + 2\pi$$

Let $z = Y(t) - x(t)$. Then

$$\frac{dz}{dt} = f(t, z + x) - f(t, x) \quad \text{and} \quad z(0) = 2\pi$$

But, $z(t) = 2\pi$ is the unique solution of this problem! Therefore, $Y(t) = x(t) + 2\pi$, and so

$$\phi(x + 2\pi) = \phi(x) + 2\pi$$

for all x. This is referred to as the *circle mapping property*.

Iteration of Circle Mappings

Let x_{n+1} denote the image of x_n under Poincare's mapping:

$$x_{n+1} = \phi(x_n)$$

In particular, given a point x_0, the dynamics of the solution $x(t)$ are described by the sequence $\{x_n\}$, which gives the values of the solution $x(t)$ each time it crosses the circle $t = 0$. It is similar to flashing a light on the solution $x(t) \bmod 2\pi$ at each time t hits an integer multiple of 2π, and so this is sometimes referred to as being the *stroboscopic method* [2.25].

Because of the circle mapping property, it is necessary to consider only the iterates of ϕ modulo 2π. In fact, if we define a sequence y_n to be the residues of the sequence x_n modulo 2π (i.e., $x_n = 2\pi M_n + y_n$ for some integer M_n, where $0 \leq y_n < 2\pi$), then

$$x_{n+1} = y_{n+1} + 2\pi M_{n+1} = \phi(x_n) = \phi(y_n + 2\pi M_n) = \phi(y_n) + 2\pi M_n$$

so

$$y_{n+1} = \phi(y_n) \bmod 2\pi.$$

For example, consider the differential equation

$$\dot{x} = \alpha$$

where α is a constant. The solutions are

$$x(t) = \alpha t + x(0)$$

Therefore,

$$\phi(x) = x + 2\pi\alpha$$

so ϕ amounts to rotation through $2\pi\alpha$ radians. This case is referred to as the Kronecker flow on the torus [2.26]. The iterations modulo 2π become

$$y_{n+1} = y_n + 2\pi\alpha \qquad (\bmod 2\pi)$$

These are plotted in Figure 2.27.

We see that if α is rational, then every orbit closes, since in that case $y_n = y_0 + 2\pi\alpha n = y_0$ when n is a multiple of the denominator in α. But, if α is irrational, every orbit is dense in the torus.

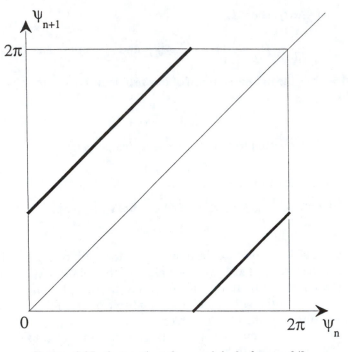

FIGURE 2.27. $\psi_{n+1} = \psi_n + 2\pi\alpha$ modulo 2π for $\alpha = 3/8$.

The *rotation number* is defined for Poincare's mapping by the formula

$$\rho = \lim_{t \to \infty} \frac{x(t)}{t} = \lim_{n \to \infty} \frac{\phi^n(\xi)}{2\pi n}$$

where ξ is the initial condition of $x(t)$. Recall that ρ does not depend on the choice of ξ (see Section 2.2.2).

For $dx/dt = \alpha$, we have

$$\lim_{n \to \infty} \frac{\phi^n(\xi)}{2\pi n} = \lim_{n \to \infty} \frac{2\pi n\alpha}{2\pi n} = \alpha$$

Therefore, we have a simple illustration of Denjoy's theorem.

The next example illustrates a use of the rotation number in computer simulations of a first-order VCON. We write the system as

$$\dot{x} = \omega + \cos_+ x \cos_+ y$$

$$\dot{y} = 1$$

Figure 2.28 shows the rotation number of this equation where $\rho = \lim_{t \to \infty} x(t)/t$. We see clearly intervals where the graph of ρ plateaus, and it is over these that x is phase-locked to y.

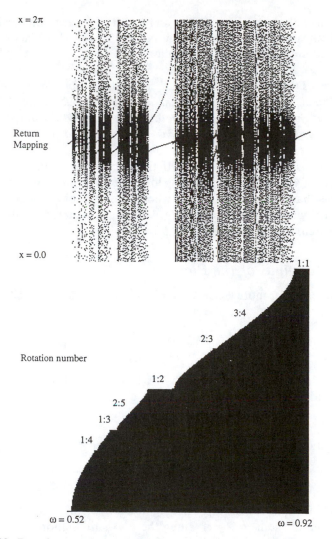

FIGURE 2.28. Rotation number and return map simulations for the VCON. Note that there is a unique fixed point for the return mapping when $\rho = 1$, two points when $\rho = 1/2$, etc.

2.5.3 *Annulus Mappings*

A variety of interesting results have been obtained by Poincare and Birkhoff for mappings of an annulus into itself [2.27]. An annular region in the plane is one that is bounded by two concentric, nonintersecting smooth arcs. Mappings of an annular region into itself occur naturally when we study the response of a nonlinear oscillation to periodic forcing. The forced oscillation often lies near the periodic solution, and so an invariant annular region is formed. Poincare's Twist Theorem can be applied to prove the existence of periodic and more complicated oscillations.

Poincare's Twist Theorem. *Suppose that* Φ *maps an annular region A in* E^2 *into itself such that* Φ *preserves area and leaves the boundaries of A invariant. Suppose that the outside boundary is rotated one way and the inside boundary the other way. Then there must be at least one fixed point for* Φ *in A.*

The proof is not presented here. If the conditions of this theorem are satisfied and if Φ is Poincare's mapping for some dynamical system, then that system must have a periodic solution whose stroboscopic iterates lie in A. In fact, the behavior of iterates of Φ are usually much richer, as in the example of this theorem discussed in Section 2.6.

2.5.4 *Mappings of the Plane*

Let us now consider more general mappings of the plane into itself. We begin with a linear iteration

$$x_{n+1} = (1 + \sigma)x_n$$

$$y_{n+1} = (1 - \tau)y_n$$

where σ and τ are real numbers with $0 < 1 - \tau < 1 < 1 + \sigma$. Following the work of Hadamard [2.28], we begin with a circle of initial data

$$x_0^2 + y_0^2 = c^2$$

where c is a constant. Then the nth iterate satisfies

$$\left[\frac{x_n}{(1 + \sigma)^n}\right]^2 + \left[\frac{y_n}{(1 - \tau)^n}\right]^2 = c^2$$

which is the formula for an ellipse having radius $c(1 + \sigma)^n$ along the x axis (its major axis) and radius $c(1 - \tau)^n$ along the y axis (its minor axis). Thus, the circle generates a sequence of ellipses that converge to the x axis.

The point $x = 0$, $y = 0$, is an equilibrium for this system. It is a hyperbolic (saddle) point since its characteristic multipliers are $1 + \sigma > 1 > 1 - \tau$.

Next, consider a nonlinear mapping

$$x_{n+1} = (1 + \sigma)x_n + f(x_n, y_n)$$

$$y_{n+1} = (1 - \tau)y_n + g(x_n, y_n)$$

where f and g are smooth functions having no linear terms in their Taylor expansions near $x = 0$, $y = 0$. We write

$$(x_{n+1}, y_{n+1}) = \Phi(x_n, y_n)$$

for this iteration. In analogy with Perron's Theorem for differential equations (Chapter 3), we expect there to be a smooth stable manifold and a smooth unstable one. This fact will not be proved here. However, we can see that this is the case near $x = 0$, $y = 0$.

First, find a function $y = Y(x)$ for x near 0 such that

$$Y = (1 - \tau)Y + g(x, Y)$$

and $Y(0) = 0$. This can be done since the following conditions are satisfied

1. $x = 0$, $Y = 0$ is a solution.
2. The function g is smooth near this point.
3. The Jacobian $\{\partial[-\tau Y + g(x, Y)]/\partial Y\}(0,0) = -\tau \neq 0$.

Therefore, the implicit function theorem ensures that there is a unique solution of this equation that satisfies $Y(0) = 0$. The solution is smooth, and it can be expanded in a Taylor polynomial about $x = 0$.

In fact, the set

$$U = \{(x, y): y = Y(x)\}$$

is the unstable manifold for this system, and any point beginning on it [say, $y_0 = Y(x_0)$] generates a sequence (x_n, y_n) that diverges from $(0,0)$ as n increases.

Similarly, we can find a function $x = X(y)$ that solves the problem

$$X = (1 + \sigma)X + f(X, y)$$

for y near 0. The set $S = \{(x, y): x = X(y)\}$ is the stable manifold for this problem.

Homoclinic Points

The stable and unstable manifolds can eventually cross, and, when this happens, the consequences are dramatic. Suppose that there is a point $R =$

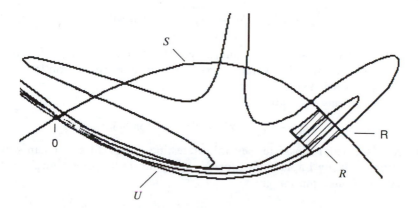

FIGURE 2.29. S and U for a transverse homoclinic point. Infinitely many periodic orbits hit the rectangle R. In addition, given any sequence of 0's and 1's there are two sets O and I in R and an orbit that hits these two sets in the order listed in the given sequence. [See 2.29]

$(\xi, \eta) \in S \cap U$. Then, $\Phi^n(R) \to (0,0)$ as $n \to \pm\infty$. It can also happen that the manifolds S and U cross transversally at R; that is, the tangents to these two manifolds at R are not collinear, and we say that R is a *transversal homoclinic point*. In this case, the loop in Figure 2.29 on the manifold U between R and $\Phi(R)$ must approach S as $n \to \infty$. It becomes terribly wrapped under iterations of Φ.

In particular, consider the rectangular region **R** in Figure 2.29. It is known that **R** contains an invariant set that has infinitely many periodic solutions in it, as well as solutions that remain in **R**, but never repeat. These are aperiodic oscillations. In fact, there are two subsets of **R**, say labeled 0 and 1, such that given any sequence of 1s and 0s, there is a solution that hits the set 0 and the set 1 in the order specified by this sequence. This also occurs in forced oscillators, as is shown later [see 2.29].

2.6   Oscillations of Equations with a Time Delay

Equations with time delays can often be analyzed using iteration methods. Consider a system of differential-difference equations

$$\dot{x} = f[x(t), x(t - \sigma)]$$

for $x, f \in E^N$, and where $\sigma > 0$ is a fixed time delay. It is shown in Ref. 2.30 that, if f is a smooth function of its arguments and if x is specified on some interval of length σ, say

$$x(t) = \psi(t) \qquad \text{for } -\sigma \le t \le 0$$

then there is a unique solution of this system on some future interval. In fact, on the interval $0 \le t \le \sigma$, this is just the system of ordinary differential equations

$$\frac{dx}{dt} = f[x(t), \psi(t - \sigma)], \qquad x(0) = \psi(0)$$

and all the usual existence and uniqueness theory for ordinary differential equations carries over. If a solution exists, the process can be continued by marching forward in steps of size σ.

The linear delay equation

$$\dot{x} = -\alpha x(t - 1), \qquad x(t) = \psi(t) \qquad \text{for } -1 \le t \le 0$$

where α is a scalar illustrates several interesting things. First, we can solve this problem using Laplace transforms. Applying the Laplace transform to both sides of this equation gives

$$p x^*(p) - x(0) = -\alpha \exp(-p) x(p) + \int_0^1 \exp[-p(s + 1)] \psi(s - 1)\, ds$$

Therefore, the characteristic equation is

$$p + \alpha \exp(-p) = 0$$

since this equation describes the poles of the problem's transfer function [see Chapter 1]. Reference 2.30 further shows that there are countably many complex number solutions to this equation, say $\{p_m\}$, and that the solution $x(t)$ can be written as the series

$$x(t) = \sum_{n=0}^{\infty} B_n \exp(p_n t)$$

The complex values for p that solve the characteristic equation have the form $p = \rho + i\tau$, and so separating real and imaginary parts in the characteristic equation gives an equivalent system

$$\rho = -\alpha e^{-\rho} \cos \tau$$

$$\tau = \alpha e^{-\rho} \sin \tau$$

When $\alpha = \pi/2$, there are two imaginary characteristic values: namely, $\rho = 0$, $\tau = \pm\pi/2$. For that value of α, the equation has periodic solutions. When $\alpha < \pi/2$, then Re $\rho < 0$, and so $x(t) \to 0$ as $t \to \infty$. Finally, when $\alpha > \pi/2$, there is a characteristic root having $\rho > 0$, so solutions grow as $t \to \infty$.

It is appropriate to think of a differential-difference equation as being one for functions that are in an infinite dimensional space that is spanned by the functions $\{\exp(p_n t)\}$. This observation leads off into mathematical results that have been effectively used to study nonlinear parabolic partial differential equations [2.31].

Note that setting $\tau = \alpha t$ and $y(\tau) = x(\tau/\alpha)$ changes the example into the equation

$$dy/d\tau = -y(\tau - \alpha)$$

Therefore, increasing the value of α has the effect of increasing the delay. The Laplace transform calculations show that oscillations can appear as the delay increases through $\alpha = \pi/2$.

2.6.1   Linear Spline Approximations

A useful technique for approximating differential-difference equations of the form

$$\frac{dx}{dt} = f[x(t), x(t - \sigma)]$$

with the goal of solving them numerically, was derived in Ref. 2.32. This approach also gives useful insight to solutions of these equations. The idea is to take the delay interval of length σ and divide it up into N equal parts. At each point of the partition, say t_j, we define $x(t_j)$ to be a *node* of a *linear spline*. We will approximate the function $x(t)$ over this interval by taking a piecewise linear function that agrees with x at the nodes t_j. The approximating function

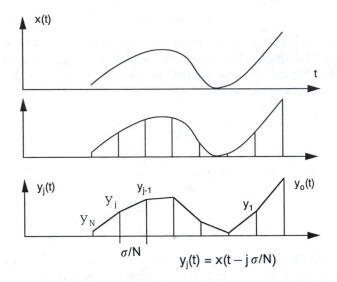

FIGURE 2.30. (Top) $x(t)$. (Middle) An equal partition of the t axis on which the linear spline approximation is based. (Bottom) The linear spline approximation to $x(t)$.

is described by a vector $(y_0, \cdots, y_N)$: Let

$$y_j(t) = x(t - j\sigma/N)$$

for the $N + 1$ nodes $j = 0, \cdots, N$. In particular, $y_N(t) = x(t - \sigma)$. Between nodes, y_j is a linear function. We interpret differentiation in the equation to be from the right.

With this notation, we see that the derivative of $y_j(t)$ is approximately

$$\dot{y}_j \approx \frac{y_{j-1} - y_j}{\sigma/N}$$

Using the linear spline approximation, we obtain the system of ordinary differential equations (containing no time delay)

$$\dot{y}_0 = f(y_0, y_N)$$

$$\dot{y}_j = \frac{N}{\sigma}(y_{j-1} - y_j)$$

for $j = 1, \cdots, N$, for the components of the vector $\mathbf{y}(t)$. With these notations, we have the following theorem.

Theorem on Linear Spline Approximation. *Under the conditions listed above,*

$$x(t) = y_N(t) + O(1/N)$$

as $N \to \infty$, uniformly for t on any finite set over which $x(t)$ exists.

The proof of this result is given in Ref. 2.32.

Example. Linear Spline Solution. Consider the equation

$$\dot{x} = -\alpha x(t-1)[1 - x^2(t)]$$

The linear spline approximation method with $N = 2$ gives

$$\dot{y}_0 = -\alpha y_2(t)[1 - y_0^2(t)]$$
$$\dot{y}_1 = -2(y_1 - y_0)$$
$$\dot{y}_2 = -2(y_2 - y_1)$$

This system has equilibria when $y_2 = y_1 = y_0$ for $y_0 = 0$ and $y_0 = \pm 1$. The last two are hyperbolic. Linearizing about $x = 0$ gives the same equation as in the preceding example. For $\alpha < \pi/2$, it is stable. For $\alpha > \pi/2$, there appears a nonlinear oscillation.

2.6.2   *Special Periodic Solutions*

Consider the equation

$$\dot{x} = h[x(t)]f[x(t-1)]$$

where f and h are smooth functions. We can transform this equation by a change of variables $x \to g(z)$ into the form

$$\dot{z} = F[z(t-1)]$$

In fact,

$$g'(z)\dot{z} = f\{g[z(t-1)]\}h[g(z)]$$

so if $g' = h(g)$, we have that $F(z) = f[g(z)]$. We see that if $h(x) = 1 - x^2$ as in Section 2.6.1, then $g = \tanh z$.

Thus, we consider the equation

$$\dot{z} = F[z(t-1)]$$

where F is a smooth function. We look for special periodic solutions of this equation that satisfy the conditions

$$z(t+2) = -z(t), \qquad z(-t) = -z(t)$$

Setting $y(t) = z(t-1)$ and $x = -z$ gives

$$\dot{x} = -F(y)$$
$$\dot{y} = F(x)$$

This Hamiltonian system was derived in Ref. 2.33.

An interesting feature of this system has been pointed out [2.33]. Suppose that we try to solve for special periodic solutions using a semiimplicit Euler

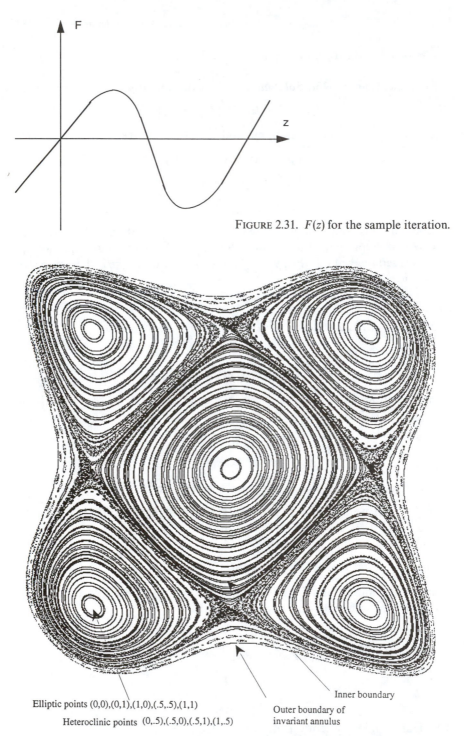

FIGURE 2.31. $F(z)$ for the sample iteration.

Elliptic points (0,0),(0,1),(1,0),(.5,.5),(1,1)

Heteroclinic points (0,.5),(.5,0),(.5,1),(1,.5)

Inner boundary

Outer boundary of
invariant annulus

FIGURE 2.32. An invariant annulus is shown in iterations of the area preserving mapping described in the text. Note that there are five elliptic points and four hyperbolic (heteroclinic) points.

numerical algorithm, say

$$y_{n+1} = y_n + ahF(x_n)$$

$$x_{n+1} = x_n - ahF(y_{n+1})$$

This defines a mapping of the (x, y) plane into itself, say $T(x_n, y_n)$. The Jacobian of this system is given by

$$\det \begin{bmatrix} 1 & ahF'(x_n) \\ -ahF'(y_{n+1}) & 1 - a^2h^2F'(y_{n+1})F'(x_n) \end{bmatrix} = 1$$

Therefore, T is an area preserving transformation of the plane.

Suppose that $F(z)$ has the form shown in Figure 2.31. Then the equilibria are determined from the equations

$$F(x^*) = F(y^*) = 0$$

and their stability is determined using the eigenvalues of the matrix

$$\begin{bmatrix} 0 & -F'(y^*) \\ F'(x^*) & 0 \end{bmatrix}$$

Moreover, for F as shown in Figure 2.31, this iteration has an invariant annulus. Figure 2.32 shows a numerical simulation of this iteration.

Poincare's Twist Theorem (Section 2.5.8) shows that there is at least one fixed point within this annulus. However, the actual behavior of the iteration is much more complicated.

2.7 Exercises

2.1. Reproduce the phase plane portraits for the four cases of a two-dimensional linear system with constant coefficients, as depicted in Figures 2.1–2.4.

2.2. The Poincare-Bendixson Theorem indicates that there are three possible evolutions for a bounded trajectory of a two-dimensional autonomous system: It approaches a stable equilibrium, a stable periodic orbit, or a stable necklace. Construct an example of each of these cases.

2.3. a. FitzHugh and Nagumo's model arises in studies of tunnel diodes and neurons:

$$dv/dt = w - f(v)$$

$$dw/dt = v - w$$

where $f(v) = v(v - 1)(v - a)$ and a is a fixed parameter $0 \leq a \leq 1$. Show that this system has a unique periodic solution using the results for Lienard's equation.

b*. Consider the general Lienard equation

$$\frac{d^2x}{dt^2} + g(x)\frac{dx}{dt} + U_x(x) = 0$$

Suppose that U and g are smooth functions (at least continuously differentiable), that g is an even function of x [that is, $g(-x) = g(x)$ for all x], U is an odd function [that is, $U(-x) = -U(x)$ for all x], and that $U(\pm\infty) = \infty$. Also, suppose that there is a number α such that

$$G(x) = \int_0^x g(s)\,ds$$

is negative for $x < \alpha$ and positive monotone for $x > \alpha$ with $G(\infty) = \infty$. Show that there is a unique nontrivial periodic solution. Moreover, show that any nontrivial solution approaches this solution as $t \to \infty$. (*Hint*: Consider the function $E(x, dx/dt) = (dx/dt)^2/2 + U(x)$ in the proof in Section 2.1.3)

2.4. Consider the differential equations

$$\frac{dx}{dt} = 1, \qquad \frac{dy}{dt} = \alpha$$

for (x, y) on the torus $T = \{(x, y): 0 \le x, y < 2\pi\}$
a. Show that the rotation number of this system is $\rho = 1/\alpha$. Describe solutions when α is a rational number. What happens to solutions when α is irrational? Why? Discuss the computability in these two cases by dividing the torus up into small squares for which each is darkened and remains so when a trajectory hits it. Corroborate your discussion with a computer simulation illustrating each case.
b. If $\alpha = 1/29.6$, plot $(\cos x(t) \cos y(t))_+$ over a toroidal patch $0 \le x, y \le 2\pi$. If t denotes solar time in days, then the pair (x, y) simultaneously record the solar and lunar times.
c. Use a computer simulation to calculate the rotation number of the system

$$\frac{dx}{dt} = 1 + \lambda + (1 - \lambda)\sin(2x - y) + \lambda\sin(x - 2y)$$

$$\frac{dy}{dt} = 2 - \lambda$$

for $0 \le \lambda \le 2$. Plot ρ versus λ for each choice you make of λ (see Figure 2.8).
d. Consider the equations

$$dx/dt = p \qquad dy/dt = q$$

where p and q are integers and $x(0)$ and $y(0)$ are given, say $x(0) = 1.0$, $y(0) = 1.0$. Show that $\rho = p/q$. The solution of this system is a *toroidal knot* having winding numbers p and q. Plot the solution of this equation in the four cases

$$\text{i. } p = 1 \qquad q = 2$$

$$\text{ii. } p = 1 \qquad q = 3$$

$$\text{iii. } p = 2 \qquad q = 3$$

$$\text{iv. } p = 5 \qquad q = 6$$

Deduce that for a general system on the torus as considered by Denjoy (section 2.2.2) if the rotation number is rational (so all solutions approach a periodic solution whose rotation number is p/q), then there is a torus knot that is a stable oscillation of the system.

2.5. a. Consider the three differential equations

$$\frac{dx}{dt} = a + \sin(x - y)$$

$$\frac{dy}{dt} = b + \sin(y - z)$$

$$\frac{dz}{dt} = c + \sin(z - x)$$

where $a, b, c > 0$ and $a + b + c > 3$
Reduce this system to one on a torus using the projection method in Section 2.2.3. (Hint: $x + y + z$ is a time like variable)
What is $\lim_{t \to \infty} x(t) : y(t) : z(t)$ if $a = 2$, $b = 2$, and $c = 2$?

b*. The common ratios in part b can be plotted using areal coordinates: Let $p = x/(x + y + z)$, $q = y/(x + y + z)$, $r = z/(x + y + z)$. If $a > 1$, $b > 1$, and $c > 1$, then these three numbers are nonnegative and add to 1.0. Therefore, the point (p, q, r) can be identified with one lying on an equilateral triangle in the positive orthant of E^3. Plot these numbers for $t = 100\pi$ and for various choices of a, b, and c using a computer (see Fig. 7.4).

2.6. a. Plot the phase portrait of solutions to the conservative equation

$$\frac{d^2x}{dt^2} + U_x(x) = 0$$

corresponding to each of the potential functions

i. $U(x) = x^2/2$

ii. $U(x) = a(\sin x/2)^2$

b. Show that in case i, all solutions have period 2π.

c*. Calculate the period of oscillations in case ii near stable equilibria. Describe the period of oscillations of bound states as a function of amplitude by plotting the energy E versus the period of the oscillation having this energy. Use this calculation to discuss the solution of the two-point boundary-value problem

$$\frac{d^2x}{dt^2} + U_x(x) = 0, \qquad x(0) = 0, \qquad x(\tau) = 0$$

For what values of τ is this solvable? How many solutions are there?

2.7. a. In Section 2.3.5 we derived the transformation from cartesian coordinates to phase-amplitude coordinates for Hamiltonian systems: $p = p(a, \xi)$, $q = q(a, \xi)$. Calculate the Jacobian of this transformation. Carry out the full transformation in the case $U(x) = x^2/2$. Derive the change of variables and use it to solve the equation

$$\frac{d^2x}{dt^2} + U_x(x) = 0.$$

b*. Carry out the same program when $U(x) = (\sin x/2)^2$. (*Hint*: See Ref. 2.7).

2.8. a. Consider the Hamiltonian $H(p,q) = \sum (p_i^2 + q_i^2)/2 + \varepsilon H_1(p,q)$ where ε is a small number and $H_1(p,q)$ is a smooth (differentiable) function. Write the resulting Hamiltonian system in phase-amplitude coordinates. (*Hint*: Set $dq_j/dt + i\omega_j q_j = y_j \exp ix_j$.)

b. A collection of N bodies, say located at points $x_1, \cdots, x_N$, in E^3 and having masses $m_1, \cdots, m_N$, satisfy the equations

$$m_j \frac{d^2 x_j}{dt^2} = -\frac{\partial U}{\partial x_j}$$

for $j = 1, \cdots, N$, where

$$U = \sum \frac{m_k m_j}{|x_k - x_j|}$$

and the sum is taken over all indices j and k for which $j \neq k$. Carefully write out the resulting Hamiltonian system.

c. The two-body problem results in part b when $N = 2$. Use a computer simulation to plot solutions to the two-body problem.

2.9. Use the Hamilton-Jacobi method to solve the initial value problems

$$\frac{\partial \phi}{\partial t} + x \left(\frac{\partial \phi}{\partial x} \right) = 0$$

$$\phi(0, x) = x, \qquad \text{where } -\infty < x < \infty.$$

and

$$\frac{\partial \phi}{\partial t} + (x, \operatorname{div} \phi) = 0,$$

$$\phi(0, x) = g(x)$$

where x is in E^N and (a, b) denotes the dot product of two vectors a and b.

2.10. Let $U(x) = x^4/4 - ax^2/2 + bx + c$.

a. Describe the solutions of the differential equation

$$\frac{d^2 x}{dt^2} + U_x(x) = 0$$

for all choices of a, b, and c, using the methods of Sections 2.3.2 and 2.4.3.

b. Compute and plot the solutions of van der Pol's equation and the lock-washer oscillators in Section 2.4.3 for $\varepsilon = 0.1$. Also, identify the underlying the cusp surface.

2.11. Let f be a single-valued function mapping the set $S = \{1, \cdots, N\}$ into itself.

a. Show that the iterates of f, say f^j, define a semigroup using the operation of function composition: That is, show that $f^i f^j = f^{i+j}$.

b. Show that the spectrum of the graph transition matrix for f consists of zeros and roots of unity. Relate the corresponding eigenvectors to the invariant sets of f.

c. Perform a Monte Carlo simulation of the matrix M associated with the function

$$g(x) = 3.54x(1.0 - x)$$

as in Section 2.5.1.

2.12. a. Show that the function ϕ defined in Section 2.5.2. defines a circle mapping. In particular, show that the solutions $x(t)$ and $\xi(t)$ of the differential equation

$$dx/dt = f(t, x)$$

where f is doubly periodic and where the initial data satisfy

$$\xi(0) \text{ is arbitrary and } x(0) = \xi(0) + 2\pi$$

satisfy

$$x(t) = \xi(t) + 2\pi$$

for all t.

b. Show that the residues of $x(2\pi n)$ modulo 2π satisfy the iteration

$$y_{n+1} = \phi(y_n)(\text{mod } 2\pi).$$

Also, show how to use the sequence y_n to reconstruct the sequence x_n. (*Hint*: $x_0 - y_0 = 2\pi M$ for some integer multiple M.)

c. Show that if α is rational, then all iterates defined by

$$y_{n+1} = y_n + 2\pi\alpha$$

are periodic. Show that if α is irrational, then all iterates are dense in the circle $0 \le y < 2\pi$.

d. Consider the mapping $r\exp(i\theta) \to 2r\exp(2i\theta)/(1 + r)$. Show that the circle $r = 1$ is invariant for this mapping; in fact, show that all points in the plane (except $r = 0$) approach this circle under iterations of this mapping. Show that there are infinitely many distinct periodic solutions on the circle. [*Hint*: The mapping restricted to the circle $r = 1$ defines the circle mapping $\theta_{n+1} = 2\theta_n(\text{mod } 2\pi)$.] The set $r = 1$ is a strange attractor for this system.

2.13. Reproduce Figure 2.26.

2.14. a. Routh-Hurwitz Criterion for stability. Show that if all of the coefficients of a polynomial are positive, then all of its roots must have negative real parts [see 2.31].

b*. The differential-difference equation $dx/dt = -ax(t - 1)$ can be solved using Laplace's transform. Show that the characteristic equation in this case is $p = -ae^{-p}$. Consider the roots for p of the function $p + ae^{-p}$. Show that $\sin(\pi t/2)$ solves the problem when $a = \pi/2$. Show that for $a < \pi/2$ all roots have negative real parts. Describe how the set of roots changes as a increases beyond $\pi/2$. (*Hint*: $pe^p + a = 0$. Let $p = \sigma + i\tau$, and derive equations relating a, σ, and τ.)

2.15. a. Describe all solutions of the Hamiltonian system

$$dz/dt = -F(y)$$

$$dy/dt = F(z)$$

where $F(y) = y(1 - y)(y - a)$ and a is a fixed number $0 < a < 1$. [*Hint*: Show that $G(z) + G(y)$ is constant along solutions, where $G'(z) = F(z)$.]

b. Discretize this model using the semiimplicit method described in Section 2.6.2. Show that the trransformation T defined by this iteration satisfies $\det(T) = 1$. Show that this implies that the transformation is area preserving.

3
Stability of Nonlinear Systems

A mechanical or an electrical device can be constructed to a level of accuracy that is restricted by technical or economic constraints. What happens to the output if the construction is a little off specifications? Does output remain near design values? How sensitive is the design to variations in construction parameters?

Stability theory gives some answers to these and similar questions. The ultimate concept of stability developed here is *stability under persistent disturbances*. Much of our later work in perturbation theory is based on this idea. However, *linear stability theory* is the most convenient concept to apply. It is based on a study of small deviations from a design state, and so it reduces the problem to a linear problem for small deviations. The methods of Chapter 1 can be used once this is done.

Sometimes the linear analysis carries little or no useful information. For example, if large deviations must be considered, then the linear analysis is unreliable, or the small deviations problem might have neutral stability properties that do not carry over directly to the full model. In some cases, the linear problem is unstable but nonlinearities capture growing modes with the result being that the design output is still stable. In these cases, other methods based on energy considerations are useful, and Liapunov's theory is the one developed and used here.

3.1   Desirable Stability Properties of Nonlinear Systems

Consider a system of differential equations of the form

$$dx/dt = f(t, x)$$

where $x, f \in E^N$, the components of f are smooth functions of their arguments, annd as before, t is a real variable. Assume throughout this section that the components of f are at least continuous in t and twice continuously differentiable with respect to the components of x.

Suppose that there is a solution of this equation, say $x = \phi(t)$, that exists for all future times, say $t_0 \leq t < \infty$. Deviations from ϕ are described by the variable $y = x - \phi$. With this the equation becomes

$$dy/dt = F(t, y)$$

where $F(t, y) = f[t, y + \phi(t)] - f[t, \phi(t)]$. This new equation has $y = 0$ as a solution. In this way, the study of a solution $x = \phi(t)$ is reduced to the study of the static state $y = 0$ of an associated problem. For the convenience of notation, we return to the original notation and suppose that $f(t, 0) = 0$ for all $t, t_0 \leq t < \infty$.

The question asked in this section is: How do solutions beginning near $x = 0$ behave? We hope that they stay near $x = 0$ for all time, or even approach it as $t \to \infty$, and the methods developed in this section provide some useful ways to determine this.

We begin with a series of definitions. Let $x(t, t_0, x_0)$ denote the solution of the system

$$dx/dt = f(t, x)$$

that satisfies the initial conditions

$$x(t_0, t_0, x_0) = x_0.$$

Since $f(t, 0) = 0$, we see that $x(t, t_0, 0) \equiv 0$.

1. The solution $x \equiv 0$ is said to be *stable* if given any tolerance $\varepsilon > 0$ and any initial time t_0, there is a constraint $\delta(\varepsilon, t_0) > 0$ such that $|x_0| < \delta$ implies that $x(t, t_0, x_0)$ exists for $t_0 \leq t < \infty$ and it satisfies

$$|x(t, t_0, x_0)| < \varepsilon$$

   for all $t \geq t_0$. Thus, a solution will stay near $x = 0$ if it starts nearby.

2. The solution $x \equiv 0$ is said to be *asymptotically stable* if it is stable and if there is a constraint $\delta_1(\varepsilon, t_0) > 0$ such that $|x_0| < \delta_1$ implies that

$$|x(t, t_0, x_0)| \to 0$$

   as $t \to \infty$. Thus, $x = 0$ is stable and solutions starting near it will approach it.

3. The solution $x \equiv 0$ is said to be *uniformly asymptotically stable* (UAS) if it is stable with its constraint δ being independent of t_0 and if given a tolerance $\varepsilon > 0$ there is a number $T(\varepsilon)$ such that $t - t_0 > T(\varepsilon)$ implies that

$$|x(t, t_0, x_0)| < \varepsilon$$

   Thus, x approaches zero as $t - t_0 \to \infty$, uniformly in t_0 and in x_0.

4. The solution $x \equiv 0$ is said to be *exponentially asymptotically stable* (EAS) if there are constants δ, K, and α, all positive, such that $|x_0| < \delta$ implies that

$$|x(t, t_0, x_0)| \leq Ke^{-\alpha(t-t_0)}|x_0|$$

   for all $t \geq t_0$.

5. The solution $x \equiv 0$ is said to be *stable under persistent disturbances* (SPD) if given a tolerance ε there are constraints δ_1 and δ_2 such that (a) for any initial state η satisfying $|\eta| < \delta_1$ and (b) for any perturbing function $g(t, x)$ that is continuously differentiable with respect to the components of x, Lebesgue integrable in t, and satisfying $|g(t, x)| < \delta_2$ for $t_0 \leq t < \infty$ and for x near $x = 0$, then the solution of the perturbed equation

$$dy/dt = f(t, y) + g(t, y)$$

that satisfies the initial condition

$$y(t_0) = \eta$$

meets the tolerance

$$|y(t)| < \varepsilon$$

for $t_0 \leq t < \infty$.

Stability under persistent disturbances, sometimes called *total stability* [see 3.1], plays a central role in perturbation theory as shown in later chapters. It states that adding a small, but possibly very erratic, perturbation to the original system might change the solutions, but it does not change them much: they still remain near $x = 0$. The condition that g be measurable in t is quite a mild one, but it can be relaxed still further [3.2].

The *domain of attraction* of an asymptotically stable state is the set of all initial values x_0 such that $x(t, t_0, x_0)$ approaches the state as $t \to \infty$.

These definitions are used in many contexts. Stability and asymptotical stability are not strong properties, in the sense that they are not preserved under small changes in the system. However, EAS, UAS, and SPD are quite strong ideas, and will be developed further in this chapter. Note though that asymptotical stability in a system that is time invariant or periodic in t implies UAS. Of course, a solution that is EAS is also UAS [see 3.3]. First, we study systems that are exponentially asymptotically stable.

3.2   Linear Stability Theorem

Taylor's expansion of $f(t, x)$ about $x = 0$ can be used to derive a linear problem for small deviations from $x = 0$. Since $f(t, 0) = 0$, expanding f about $x = 0$ gives the formula

$$f(t, x) = A(t)x + G(t, x)$$

where A is the Jacobian matrix having components

$$A_{i,j}(t) = \frac{\partial f_i}{\partial x_j}(t, 0)$$

for $i, j = 1, \cdots, N$. G is the remainder in Taylor's formula, so for each t there is a constant K_1 such that

$$|G(t, x)| \leq K_1 |x|^2$$

Denote this property by $G(t, x) = O(|x|^2)$ as $x \to 0$ for each t; the notation $O(|x|^2)$ is explained fully in Chapter 5. Now, suppose that this estimate holds uniformly for $t_0 \leq t < \infty$ (i.e., K_1 does not depend on t). Of course, if f is a linear function of x, then we have $G = 0$.

Solutions of the full system near $x = 0$ can be studied using the linear problem and ignoring G when the linear problem is EAS. Let $\Phi(t)$ be a fundamental solution of this system:

$$\frac{d\Phi}{dt} = A(t)\Phi \qquad \Phi(t_0) = \text{identity}$$

The connection between the linear problem and the full nonlinear problem can be established using the variation of constants formula and treating G as though it were known. As in Section 1.6, the result is that the differential equation

$$\frac{dx}{dt} = A(t)x + G(t, x)$$

is equivalent to the integral equation

$$x(t) = \Phi(t)x(t_0) + \int_{t_0}^{t} \Phi(t)\Phi^{-1}(s)G[s, x(s)] \, ds$$

This is an important step in many studies of nonlinear problems. The matrix Φ carries information about the behavior of solutions to the linear part of the problem, and the function G carries information about the problem's nonlinearity.

We show next that when the linear problem is exponentially asymptotically stable, then so is the nonlinear problem. Let us suppose that the linear part of the problem is EAS:

Hypothesis H. There are constants K and α, both positive, such that

$$|\Phi(t)\Phi^{-1}(s)| \leq K \exp[-\alpha(t - s)]$$

for all $t_0 \leq s \leq t < \infty$.

For example, this condition is satisfied if A is a constant matrix and all its eigenvalues have negative real parts. There are other linear systems that satisfy condition H as well (see Sections 1.4 and 3.3.4).

With this condition, we can state and prove the following useful theorem:

Linear Stability Theorem. [*See Refs. 3.2 and 3.4–3.6.*] *Let condition H hold, and suppose that* $G(t, x) = O(|x|^2)$ *as* $|x| \to 0$, *uniformly for* $t_0 \leq t < \infty$. *Then* *there is a number $\delta_0 > 0$ such that, if $|x(t_0)| < \delta_0$, then the solution emerging from this initial state approaches $x = 0$ as $t \to \infty$. In fact, there is a positive constant α_0 such that*

$$|x(t)| \le K|x(t_0)| \exp[-\alpha_0(t - t_0)]$$

for all $t \ge t_0$.

We conclude from this estimate that the solution $x = 0$ of the full non-linear problem is exponentially asymptotically stable. The proof of this theorem rests on Gronwall's inequality, described below, which is closely related to mathematical induction.

3.2.1   *Gronwall's Inequality*

Let $u(t)$, $v(t)$, and $w(t)$ be continuous (scalar) functions defined for $a \le t \le b$. Suppose that $w(t) > 0$ and that $u(t)$ satisfies the inequality

$$u(t) \le v(t) + \int_a^t w(s)u(s)\,ds$$

for all $a \le t \le b$. Then we can estimate $u(t)$ in terms of v and w; namely,

$$u(t) \le v(t) + \int_a^t w(s)v(s) \exp\left[\int_s^t w(s')\,ds'\right] ds$$

for all $a \le t \le b$.

Proof. The proof of this result follows from setting

$$R(t) = \int_a^t w(s)u(s)\,ds$$

Then $R' = wu \le wv + wR$, or

$$\frac{d}{dt}\left\{R(t)\exp\left[-\int_a^t w(s)\,ds\right]\right\} \le \exp\left[-\int_a^t w(s)\,ds\right]w(t)v(t)$$

Integrating this gives the result.   ∎

3.2.2   *Proof of the Linear Stability Theorem*

The integral equation

$$x(t) = \Phi(t)x(t_0) + \int_{t_0}^t \Phi(t)\Phi^{-1}(s)G[s, x(s)]\,ds$$

has a unique solution on some interval, say $t_0 \le t \le t_0 + \delta^*$, and it can be continued as long as the solution remains bounded [3.2]. Taking norms of both sides of this equation and using condition H gives the result

$$|x(t)| \le K \exp[-\alpha(t - t_0)]\,|x(t_0)| + \int_{t_0}^{t} K \exp[-\alpha(t - s)]\,|G[s, x(s)]|\,ds$$

If $|x|$ remains less than δ/K_1 over an interval $t_0 \le t \le T$ and δ is sufficiently small, then we have $|G(t, x)| \le \delta|x|$. Setting $u(t) = |x(t)|\,e^{\alpha t}$, we have

$$u(t) \le K \exp(\alpha t_0)|x(t_0)| + \delta K \int_{t_0}^{t} u(s)\,ds$$

for $t_0 \le t \le T$. Gronwall's inequality with $v = K \exp(\alpha t_0)|x(t_0)|$ and $w = \delta K$ shows that

$$u(t) \le K \exp(\alpha t_0)|x(t_0)| \exp[\delta K(t - t_0)]$$

Therefore,

$$|x(t)| \le K|x(t_0)| \exp[-(\alpha - \delta K)(t - t_0)]$$

If we had chosen δ so small that $\delta < \alpha K$ and restricted $|x(t_0)|$ so that $K|x(t_0)| < \delta/K_1$, then x remains so small on any interval $[t_0, T]$ that this estimate is justified. We conclude that the theorem is valid for $t_0 \le t < \infty$ with $\alpha_0 = \alpha - \delta K$. ∎

This powerful result shows that if the small deviation problem is exponentially asymptotically stable, then the nonlinear problem is also. Thus, small deviations die out for the full model.

Example. The Linear Stability Theorem shows that the solution $x = 0$ of the scalar equation

$$dx/dt = ax - x^3$$

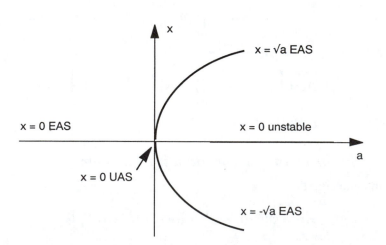

FIGURE 3.1. Pitchfork bifurcation. The static state $x = 0$ is exponentially asymptotically stable for $a < 0$, uniformly asymptotically stable for $a = 0$, and unstable for $a > 0$. As a increases through $a = 0$, two new static states bifurcate from $x = 0$. Each is exponentially asymptotically stable.

is exponentially asymptotically stable if $a < 0$. Note that not all solutions of this equation tend to zero. The static states and their stability are shown in Figure 3.1.

3.2.3 Stable and Unstable Manifolds

Consider the time invariant system

$$dx/dt = Ax + f(x)$$

where the constant matrix A has all eigenvalues lying off the imaginary axis. The eigenvalues then are either stable ones ($\mathrm{Re}\,\lambda < 0$) or unstable ones ($\mathrm{Re}\,\lambda > 0$). In this case, we say that A defines an *exponential dichotomy* [3.7] on the space E^N. This was discussed for linear problems in Chapter 1, and we assume without loss of generality that A has a block diagonal form

$$A = \mathrm{diag}(S, U)$$

where the matrix S has dimensions $m \times m$ and all of its eigenvalues are stable and the matrix U has dimensons $n \times n$ and all of its eigenvalues are unstable. Since these account for all eigenvalues of A, $m + n = N$. We break the vectors x and f into two vectors, say X and g, and Y and h, respectively, where X, $g \in E^m$ and $Y, h \in E^n$, and we rewrite the system as two equations

$$dX/dt = SX + g(X, Y)$$

$$dY/dt = UY + h(X, Y)$$

Perron rewrote this system [3.8] as a system of integral equations in a special way using the variation of constants formula. Consider the integral equations

$$X(t) = \exp(St)\xi + \int_0^t \exp[S(t - s)]g[X(s), Y(s)]\,ds$$

$$Y(t) = -\int_t^\infty \exp[U(t - s)]h[X(s), Y(s)]\,ds$$

If ξ is sufficiently small, then this system of equations has a unique solution, say $X = X(t, \xi)$, $Y = Y(t, \xi)$, that exists for $0 \le t < \infty$. The proof is based on successive approximations, quite like that used in proving the implicit function theorem in Section 4.1.2. The proof is not presented here.

Gronwall's inequality applies to this system, and it shows that

$$[X(t, \xi), Y(t, \xi)] \to (0, 0)$$

as $t \to \infty$. Let us define a set **S** by

$$\mathbf{S} = \{[\xi, Y(0, \xi)]\colon \xi \text{ is small}\}.$$

This set defines an *m-dimensional* manifold for ξ near zero, and any solution starting on it solves Perron's integral equations. Therefore, it approaches zero at an exponential rate, and **S** is referred to as being a *stable manifold*. It can be shown that any solution starting away from **S** is repelled from it at an exponential rate [3.2]. We summarize these results in the following theorem:

Perron's Theorem. *Suppose that the eigenvalues of the matrix A satisfy* $\text{Re}(\lambda) \neq 0$. *In particular, suppose that there are m eigenvalues with* $\text{Re}(\lambda) < 0$ *and n with* $\text{Re}(\lambda) > 0$ *where* $m + n = N$. *Finally, suppose that* $f(x)$ *is a smooth function for x near zero and* $f(0) = 0$. *Then there is an m-dimensional stable manifold for the system*

$$dx/dt = Ax + f(x)$$

Replacing t by $-t$ *in this system shows there is an n-dimensional stable manifold as* $t \rightarrow -\infty$. *We denote this by* **U**, *and it is referred to as being an* unstable manifold *for the full system.*

This result shows that the stable and unstable manifolds derived for linear problems in Chapter 1 can carry over to nonlinear systems. An interesting geometrical approach to stable and unstable manifolds was proposed by Hadamard as described in Section 2.5.4. These results can be extended to cases where A and f depend explicitly on t, but that is not carried out here. Finally, if some of the eigenvalues have $\text{Re}\,\lambda = 0$, then the nonlinear system can have a manifold of oscillatory or static solutions that is called a center manifold [3.9]. This is discussed in Section 1.2.2 and in Chapter 7.

3.3   Liapunov's Stability Theory

Linear stability theory enables us to determine the stability of critical points by linearizing the problem near them. However, linear theory is not useful in many problems. For example, the equation described in Figure 3.1 is

$$dx/dt = ax - x^3$$

It has $x = 0$ as a critical point, and if $a < 0$, it is asymptotically stable. However, this also happens when $a = 0$, that is, when the linear part of this equation is not exponentially asymptotically stable. Liapunov's theory helps resolve problems like this.

3.3.1   *Liapunov's Functions*

Liapunov generalized the idea of a gradient system by observing certain features that can be mimicked. He derived convenient functions without reference to the physics of the problem. These are now called *Liapunov's functions* although sometimes they are still referred to as energy functions even when they are not related to a physical energy. These can be used to study stability of nonlinear systems of differential equations. The following definitions are facilitated by using *comparison functions* that are continuous (scalar) functions, say $\mathbf{a}(u)$, defined for $u \geq 0$, monotone increasing and satisfying $\mathbf{a}(0) = 0$.

A Liapunov function for the system

$$dx/dt = f(t, x)$$

is any smooth scalar-valued function

$$V(t, x): [0, \infty) x E^N \to E^1$$

for which there are comparison functions $\mathbf{a}$, $\mathbf{b}$, and $\mathbf{c}$ such that

1. $V(t, 0) = 0$ for all $t \geq 0$.
2. $\mathbf{a}(|x|) \leq V(t, x) \leq \mathbf{b}(|x|)$ for $t \geq 0$ and x near 0.
3. $\partial V/\partial t + \nabla V \cdot f(t, x) \leq -\mathbf{c}(|x|)$.

We define a comparison function $\mathbf{d}$ by the formula

$$\mathbf{d}(u) = \mathbf{c}[\mathbf{b}^{-1}(u)]$$

for later use.

The first condition states that V vanishes at the system's static state $x = 0$. The second comparisons show that if $V \to 0$, then $|x| \to 0$, and conversely. The third condition states that V is decreasing along solutions since the left-hand side of this inequality is simply the derivative of V along solutions:

$$dV[t, x(t)]/dt$$

where $x(t)$ is a solution of the system.

The definition of Liapunov's function used here is more restrictive than necessary for many results, but it is convenient for our limited goals. Further details of these definitions and applications of them can be found in Ref. 3.1, but the following examples illustrate the basic arguments of the theory.

3.3.2 *UAS of Time Invariant Systems*

Consider the time invariant system

$$dx/dt = f(x)$$

where $f(0) = 0$ and the components of f are smooth functions of the components of x for x near zero. We have the following result.

Theorem. *If there is a Liapunov function for this system, then the solution $x = 0$ is uniformly asymptotically stable.*

Proof. This result follows from two simple arguments. First, the fundamental theorem of calculus states that

$$V[x(t)] - V[x(s)] = \int_s^t \frac{dV[x(t')]}{dt} dt'$$

and so

$$V[x(t)] - V[x(s)] = \int_s^t \nabla V[x(t')] \cdot f[x(t')] \, dt' \leq -\int_s^t \mathbf{d}(V) \, dt' < 0$$

for $0 \leq s \leq t < \infty$ and for any solution with $x(0) \neq 0$. Therefore,

$$\mathbf{a}(|x(t)|) \leq V[x(t)] \leq V[x(0)] \leq \mathbf{b}[|x(0)|]$$

and so the solution $x = 0$ is stable.

If $x(t)$ starts near zero but does not approach zero as $t \to \infty$, then the fact that

$$\frac{dV[x(t)]}{dt} \leq -\mathbf{c}\{V[x(t)]\}$$

leads to the estimate

$$\int_{V[x(t)]}^{V[x(t_0)]} \frac{dV}{\mathbf{d}(V)} \geq t - t_0$$

If $x(t)$ does not approach zero, there is a number $\delta > 0$ such that $V[x(t)] \geq \delta$ for all large t. Therefore,

$$\text{Constant} \geq \int_{V[x(t)]}^{V[x(t_0)]} \frac{dV}{\mathbf{d}(V)} \geq t - t_0$$

which eventually contradicts itself. This shows that $x(t) \to 0$, and it follows that $x = 0$ is uniformly asymptotically stable.   ∎

In Section 3.3, we see that the converse of this theorem is also valid. Because of this, the idea of uniform asymptotic stability is quite natural for nonlinear systems. The remainder of this section presents examples of Liapunov's functions and some of their uses.

3.3.3 Gradient Systems

Gradient systems are important in many applications [3.10]. These are systems whose right-hand side is the gradient of a scalar function. For example, consider the system

$$dx/dt = -\nabla F(x)$$

where $x \in E^N$ and ∇F is the gradient of a smooth scalar valued function $F(x)$. One interesting thing about such systems is that an isolated minimum of F corresponds to an asymptotically stable rest point for this system. In fact, if x^* is an isolated minimum for F, then the function

$$V(x) = F(x) - F(x^*)$$

is a Liapunov function.

This observation is the basis of a numerical algorithm, called the *gradient method*, used for finding extrema of functions of several variables. Candidates for minima are among the values of x for which the gradient of F is zero: $\nabla F(x) = 0$. We have just seen that these values of x are asymptotically stable equilibria for the dynamical system

$$dx/dt = -\nabla F(x)$$

Because of this, we can begin with an initial guess, say x_0, then solve this associated gradient system using a convenient computer package and trace the evolution of the solution. If the solution converges to a point, say x^*, then we have located a candidate for a minimum of F. The computer can also be used to test x^* for being a real minimum of F directly or by calculating a Hessian matrix. Maxima of F are minima of $-F$, so the resulting system can be solved for $t \to -\infty$ to find maxima.

Example. Let us consider the entropy function

$$F(x) = -\sum_{k=1}^{N} x_k \log x_k$$

for vectors x in the probability simplex

$$\sum_p = \{x \text{ such that } x_1 + \cdots + x_N = 1 \text{ and } x_1 \geq 0, \cdots, x_N \geq 0\}$$

Then the maxima of F can be found by solving the dynamical system

$$dx_j/dt = -\log x_j - 1$$

If we use logarithms to base N, then $1 = \log N$, so this equation is

$$dx_j/dt = -\log N x_j$$

It follows that $Nx_j \to 1$ as $t \to \infty$, and so $x(t)$ approaches a uniform distribution—one expected for maximum entropy.

The function F is called the *entropy* of the distribution x in $\sum_p$, and this calculation shows that the gradient method moves the entropy toward its maximum value which occurs at the barycenter (all $x_j = 1/N$) of the simplex $\sum_p$.

3.3.4   *Linear Time Varying Systems*

Liapunov functions can be useful in studying linear systems. For example, consider the linear system

$$dx/dt = A(t)x$$

where A is a matrix of smooth and bounded functions. We ask under what conditions on A is the function

$$V(x) = |x|^2$$

a Liapunov function for this system?

The comparison functions are $\mathbf{a}(u) = \mathbf{b}(u) = u^2$, and it only remains to evaluate the derivative of V along solutions: Since $|x|^2 = x^{tr}x$, we have that

$$\frac{d(x^{tr}x)}{dt} = x^{tr}(A^{tr} + A)x$$

where A^{tr} denotes the transpose of A and x^{tr} denotes the transpose of the column vector x, so it is a row vector. Since $A^{tr} + A$ is a symmetric matrix, it is diagonalizable [3.11]. Its spectral decomposition shows that, if all of its eigenvalues have negative real parts, then there is a positive number μ such that

$$x^{tr}Ax = x^{tr}(A + A^{tr})x/2 \leq -\mu x^{tr}x$$

for all vectors x. Thus, this function is negative definite if the eigenvalues of the symmetric part of A have negative real parts. In this case, we would take $\mathbf{c}(u) = \mu u^2$. The following example illustrates this result.

Example. We consider a system of two differential equations. As in Section 1.4, let

$$U(t) = \begin{pmatrix} \cos \omega t & -\sin \omega t \\ \sin \omega t & \cos \omega t \end{pmatrix}$$

Note that U^{tr} is the inverse of U since $U^{tr}U = I$. Consider the system of equations

$$dx/dt = C(t)x$$

where $x \in E^2$, $C(t) = U^{tr}(t)BU(t)$ and

$$B = \begin{pmatrix} a & b \\ c & d \end{pmatrix}$$

First, $C(t)$ and B have the same the eigenvalues. Therefore, the eigenvalues of $C(t)$ are constants even though C itself varies with t. Second, the derivative of $|x|^2$ along solutions of the system is

$$\frac{dx^{tr}x}{dt} = (Ux)^{tr}(B^{tr} + B)(Ux) \leq \rho |x|^2$$

where ρ is any real number that majorizes $2\mathrm{Re}\,\lambda$ for all eigenvalues λ of $B^{tr} + B$. Therefore, if the eigenvalues of the symmetric part of B are stable, then the solution $x = 0$ is exponentially asymptotically stable (see Ex. 1.3).

This important example shows that stability of a linear system with periodic coefficients cannot be determined from the eigenvalues of its coefficient matrix alone. Chapter 8 shows that, if ω is sufficiently small and B is stable, then the full system is stable.

3.3.5   Stable Invariant Sets

Liapunov's functions can also describe the stability of invariant sets. We say that a set S is *invariant* with respect to solutions of the system

$$dx/dt = f(x)$$

if any solution $x(t)$ of this system with $x(t_0) \in S$ satisfies $x(t) \in S$ for all $t \geq t_0$.

For example, if x_0 is a point for which the solution $x = x(t, x_0)$ that satisfies the initial conditions $x(0, x_0) = x_0$ remains bounded for $t \geq 0$, then we define the ω-*limit* set of this solution to be the set

$$\omega(x_0) = \{\xi \text{ in } E^N \text{ such that there is a sequence } \{t_n\}$$

$$\text{with } t_n \to \infty \text{ and } x(t_n, x_0) \to \xi\}$$

Thus, the ω-limit set consists of all points that are approached by some subsequence of values lying on the orbit starting at $x = x_0$. It is known that the ω-limit set of a solution is an invariant set; that is, if $\xi \in \omega(x_0)$, then $x(t, \xi) \in \omega(x_0)$ for all $t \geq 0$ [see 3.2].

Suppose that there is a smooth function $V(x)$: $E^N \to E^1$ that has the following properties:

1. $V(x) \geq 0$
2. $\nabla V(x) \cdot f(x) \leq 0$

Let $S = \{x \in E^N \text{ such that } \nabla V(x) \cdot f(x) = 0\}$. With these notations, we have the following result.

Theorem. *If x_0 is such that $x(t, x_0)$ remains bounded for $t \geq 0$, then $\omega(x_0) \subset S$. Thus, $x(t, x_0) \to S$ as $t \to \infty$.*

Proof. First, since V is bounded below and since it is a monotone non-increasing along solutions ($V' \leq 0$), the limit $\lim_{t \to \infty} V[x(t, x_0)]$ exists, say with value c. Second, $V(\xi) = c$ for all $\xi \in \omega(x_0)$. This follows from the fact that for any such vaue ξ there is a sequence t_n such that $x(t_n, x_0) \to \xi$ and so $V[x(t_n, x_0)] \to V(\xi) = c$, since V is a continuous function. Finally, for $\xi \in \omega(x_0)$ the solution $x(t, \xi)$ remains in $\omega(x_0)$ and since V is constant on this set, we have that

$$\frac{d}{dt} V[x(t, \xi)] = \frac{dc}{dt} = 0$$

As a result, $\nabla V(\xi) \cdot f(\xi) = 0$. Thus, $\xi \in S$. These calculations show that $\omega(x_0) \subset S$.   ∎

As a result of this theorem, we see that the set S is a global attractor for solutions.

Example. Lienard's equation. Lienard's equation has the form

$$x'' + f(x)x' + h(x) = 0$$

where f and h are smooth functions. We suppose, as in Section 2.1.3, that h behaves like a restoring force:

$$xh(x) > 0 \qquad \text{for} \qquad x \neq 0$$

and that

$$H(x) = \int_0^x h(s)\, ds$$

approaches ∞ as $x \to \infty$. But now, we suppose that $f(x) > 0$ for all values of $|x| \neq 0$.

Theorem. *With these conditions on Lienard's equation, the solution $x = 0$ is globally asymptotically stable.*

Proof. To prove this, we rewrite the equation as a first-order system:

$$dx/dt = y$$
$$dy/dt = -h(x) - f(x)y$$

and we define a function V by the formula

$$V(x, y) = y^2/2 + H(x).$$

V has the following properties:

1. It is a smooth function of x and y.
2. $V > 0$ for $(x, y) \neq 0$.
3. $dV/dt = -f(x)y^2$, which is negative for $y \neq 0$.

This is not a Liapunov function for Lienard's equation since the derivative of V along solutions of Lienard's system is not strictly negative (it vanishes whenever $y = 0$ for any x.)

Lemma. Every solution of Lienard's system is bounded and approaches the largest invariant set M in $S = \{(x, y): f(x)y^2 = 0\}$.

Proof. Since $V \to \infty$ as $|x| \to \infty$, observation 3 of the above theorem ensures that all solutions are bounded. Moreover, the previous theorem shows that all solutions approach S. It follows that the solutions must approach the largest invariant set in S. This is because the ω-limit set of any orbit is an invariant subset of S, and the largest invariant set is the union of all of the ω-limit sets. ∎

Proof of Theorem (continued). The only invariant set for Lienard's equation in S is $x = 0$, $y = 0$. Therefore, it is globally asymptotically stable. This completes the proof of the Theorem.

Example. As a final illustration of this result, we consider a slight variant of van der Pol's equation:

$$x'' + x^2 x' + x = 0$$

We let

$$dx/dt = y$$
$$dy/dt = -x - x^2 y$$

and

$$V(x, y) = (x^2 + y^2)/2$$

The derivative of this function along solutions is

$$dV/dt = -x^2 y^2$$

The theorem applies to this equation where S is the union of the x and y axes. The only invariant subset of S is $(0,0)$. Therefore, all solutions approach $x = 0$, $y = 0$ as $t \to \infty$.

3.4 Stability under Persistent Disturbances

Uniform asymptotical stability might seem to be a complicated idea, but the next theorem shows that it ensures the existence of a Liapunov function. Consider the system

$$dx/dt = f(t, x)$$

where x, $f \in E^N$, f is a differentiable function, and $f(t, 0) = 0$ for all $t \geq t_0$. Massera [3.12] proved the following result.

Massera's Inverse Theorem. *Suppose that $x = 0$ is a uniformly asymptotically stable solution of this system and that f is a continuously differentiable function of its arguments for x near $x = 0$ and for $t_0 \leq t < \infty$. Then there is a Liapunov function for this system having the three properties listed in Section 3.3.1.*

The proof of this theorem is not presented here. However, the construction of Liapunov's function in Massera's proof involves integration along trajectories of the equation, and since these are not known, the proof gives few useful hints about constructing a Liapunov function. This is one of the major problems in the theory.

Still, this result plays an important role in stability theory. Firstly, it is about the only inverse theorem known, and, secondly, it gives conditions under which we can detect stability under persistent disturbances. This is a much overlooked, but very important, result due to Malkin [see 3.13].

Consider the system

$$dx/dt = f(t, x) \tag{3.1}$$

where $f(t, 0) = 0$ for all t, f is continuously differentiable in t and in the components of x for x near 0. We refer to this as being the *unperturbed equation*. A Caratheodory perturbation of this equation has the form

$$dy/dt = f(t, y) + g(t, y) \tag{3.2}$$

Suppose that g is bounded and Lebesgue measurable with respect to t and a continuously differentiable function with respect to the components of x. Such functions are referred to here as Caratheodory functions. The general question studied is what stability conditions on system (3.1) ensure that solutions of the perturbed equation (3.2) lie near $y = 0$ for any function g that is not too large.

There are many answers to this question, but Malkin found conditions on Eq. (3.1) that ensure its static state $x = 0$ is stable under persistent disturbances. Recall that the solution $x = 0$ of Eq. (3.1) is *stable under persistent disturbances* if, given a tolerance $\varepsilon > 0$, there are numbers δ_1 and δ_2 such that $|g(t, y)| < \delta_2$ for all t and y, and $|y(0)| < \delta_1$ imply that $|y(t)| < \varepsilon$ for all t, $0 \leq t < \infty$, where $x(t)$ solves Eq. (3.1) and $y(t)$ solves Eq. (3.2). Thus, if $|g|$ is small and y starts near x, then y remains close to x for all t. The following theorem gives an answer to our question.

Malkin's Theorem. *Suppose that the solution $x = 0$ is an uniformly asymptotically stable solution of Eq. (3.1). Then it is stable under persistent disturbances.*

The proof of this result involves several steps that we do not present here in detail. However, the following arguments present the essential ideas of the proof.

Massera's theorem shows that there is a Liapunov function for the system (3.1), say $V(t, x)$. Let us calculate its derivative along solutions of Eq. (3.2):

$$\frac{dV[t, y(t)]}{dt} = \frac{\partial V}{\partial t} + \nabla V \cdot f(t, y) + \nabla V \cdot g(t, y) \leq -\mathbf{c}[V(t, y)] + O(\delta_2)$$

This is strictly negative for y outside some neighborhood of $y = 0$. Hence, by the arguments used in Section 3.2.5, we see that the solutions of Eq. (3.2) approach some neighborhood of $y = 0$ as $t \to \infty$. Specifying the radius of this

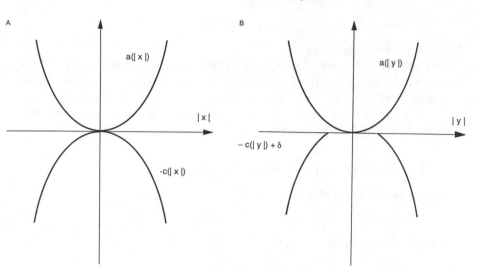

FIGURE 3.2. (a) Shown here are lower bounds for V (above) and upper bounds for dV/dt (below). The result is that when $g = 0$, $x(t) \to 0$ as $t \to \infty$. (b) Shown here are lower bounds for V (above) and upper bounds for dV/dt (below) when $g \neq 0$. Bounds for the values of dV/dt near the origin are not known, but the result is that $y(t)$ approaches a neighborhood of 0 as $t \to \infty$.

neighborhood to be ε, we adjust δ_1 and δ_2 to make the stable neighborhood sufficiently small to meet the tolerance ε. This is shown in Figure 3.2.

Note that this result does not ensure that Eq. (3.2) has a uniformly asymptotically stable solution, but only that its solutions starting near zero stay near zero. We make use of this result in some surprising ways when we study perturbation methods. Note that the conditions of the linear stability theorem imply that the solution $x = 0$ in that case is stable under persistent disturbances.

3.5 Orbital Stability of Free Oscillations

We have studied periodic solutions of conservative systems and of Lienard's equation in Chapter 2. In what senses are these solutions stable?

A periodic solution of a time invariant system cannot be asymptotically stable. To see this, consider the system

$$dx/dt = f(x)$$

where $x, f \in E^N$, and suppose that it has a (nonconstant) periodic solution, say $x = p(t)$, having a least period $T > 0$. For any number a, the function $p(t + a)$ is also a periodic solution of this system. Moreover, if a is small, it starts very near $p(t)$, but the difference

$$p(t) - p(t + a)$$

cannot approach zero as $t \to \infty$.

Because of this, a new kind of stability must be introduced; namely, *orbital stability*. Although a periodic solution of a time invariant system cannot be asymptotically stable in the sense of Liapunov (it can be stable), the *orbit* defined by the periodic solution can be. In a sense, the periodic solution can be asymptotically stable in amplitudes.

3.5.1   *Definitions of Orbital Stability*

Let $p(t)$ be a solution of least period $T(> 0)$ of the system

$$dx/dt = f(x)$$

We define the orbit of p to be the set

$$\Omega(p) = \{x = p(t): 0 \leq t \leq T\}$$

Recall that the distance between a point η and a set S is defined to be

$$d(\eta, S) = \inf\{|\eta - z|: z \in S\}$$

We say that p is *orbitally stable* if given a tolerance $\varepsilon > 0$ there is a number $\delta > 0$ such that $d[x_0, \Omega(p)] < \delta$ implies that

$$d[x(t, x_0), \Omega(p)] < \varepsilon$$

for all $t \geq 0$. Also, p is *orbitally asymptotically stable* if it is orbitally stable and solutions that start near the *orbit* approach it as $t \to \infty$. Finally, p is *orbitally stable under persistent disturbances* if given a tolerance $\varepsilon > 0$, there are numbers δ_1 and δ_2 such that for any Caratheodory function $g(t, y)$ with $|g(t, y)| < \delta_2$ and any initial point y_0 that is near the orbit, say $d[y_0, \Omega(p)] < \delta_1$, the solution $y(t, y_0)$ of

$$\frac{dy}{dt} = f(y) + g(t, y) \qquad y(0) = y_0$$

satisfies

$$d[y(t), \Omega(p)] < \varepsilon$$

for all $t \geq 0$.

3.5.2   *Examples of Orbital Stability*

A collection of excellent practice problems for studying orbital stability are presented by $\lambda\omega$-systems. Consider the system

$$\frac{dx}{dt} = \lambda(x, y)x - \omega(x, y)y$$

$$\frac{dy}{dt} = \omega(x, y)x + \lambda(x, y)y$$

where λ and ω are arbitrary functions of x and y. Polar coordinates are introduced by setting $r^2 = x^2 + y^2$ and $\theta = \tan^{-1}(y/x)$. The result is that

$$rr' = xx' + yy' = \lambda(x, y)r^2$$

and

$$r^2\theta' = xy' - yx' = \omega(x, y)r^2$$

Thus, the system becomes

$$\frac{dr}{dt} = \lambda(r\cos\theta, r\sin\theta)r$$

$$\frac{d\theta}{dt} = \omega(r\cos\theta, r\sin\theta).$$

For example, if $\omega = $ constant and $\lambda = (1 - r)$, then this system has a unique periodic orbit ($r = 1$), and it is orbitally asymptotically stable.

Recall the conservation equation

$$x'' + U_x(x) = 0$$

where the potential function U is a quartic as shown in Figure 2.3.2. There are many periodic orbits defined by this equation, and all of them are orbitally stable. However, the separatrices that begin and end at the saddle point x^* are not. These observations can be proved using the total energy

$$x'^2/2 + U(x)$$

as a measure of deviation.

Finally, recall that van der Pol's equation

$$x'' + (x^2 - 1)x' + x = 0$$

satisfies the conditions of Section 2.1.3 for Lienard's equation, so it has a unique periodic solution (other than $x = 0$). It follows from the Poincare-Bendixon Theorem that this periodic solution is orbitally asymptotically stable.

3.5.3   Orbital Stability under Persistent Disturbances

The next theorem shows how stability under persistent disturbances can result for orbits.

Theorem. *Let $p(t)$ be an orbitally asymptotically stable solution of the time invariant system*

$$dx/dt = f(x)$$

where f is a smooth function. Then $p(t)$ is orbitally stable under persistent disturbances.

The proof is not presented here [see 3.14]. However, the conditions of the theorem ensure that the orbit $\Omega(p)$ has a Liapunov function. In fact, there is a function $V(t, x)$ that is periodic in t, $V(t, x) > 0$ and $dV/dt < 0$ for x not in $\Omega(p)$. Thus, $\Omega(p)$ plays the role of S in the lemma in Section 3.2.5. The result follows from a straightforward extension of Malkin's theorem presented in the last section.

This theorem shows that the stable oscillation of Lienard's equation in Section 3.4.2 is orbitally stable under persistent disturbances.

3.5.4 Poincare's Return Mapping

Let $p(t) \in E^N$ be a periodic solution having least period $T > 0$ of the system

$$dx/dt = f(x)$$

where f is a smooth function of the components of the vector $x \in E^N$. We study the deviation of solutions from p by setting $u = x - p(t)$. Then u satisfies the equation

$$du/dt = A(t)u + g[p(t), u]$$

where A is the Jacobian matrix $A(t) = f_x[p(t)]$ and $g[p(t), u] = O(|u|^2)$ as $|u| \to 0$ uniformly for $0 \le t \le T$.

Let vectors $q_2, \cdots, q_N$ be pairwise orthogonal and orthogonal to $p'(0)$. Thus, the column vectors

$$p'(0), q_2, \cdots, q_N$$

span E^N. Let $Y(t)$ be the fundamental solution of the linear equation

$$du/dt = A(t)u$$

that satisfies the initial conditions

$$Y(0) = [p'(0), q_2, \cdots, q_N]$$

given here in terms of its columns. This equation has a periodic solution, namely $u = p'(t)$, and our choice of initial conditions ensures that $p'(t)$ is the first column of $Y(t)$.

Floquet's Theorem shows that

$$Y(t) = P(t) \exp(Lt)$$

where $P(t)$ has period T, $P(0) = Y(0)$, and

$$L = \operatorname{diag}(0, \Lambda)$$

where Λ is an $(N - 1) \times (N - 1)$ matrix. The eigenvalues of L are called the *characteristic exponents* of the system, and the eigenvalues of $\exp(LT)$ are called the *characteristic multipliers*. Note that at least one of the characteristic multipliers must have a modulus equal to 1 since the function $u = p'(t)$ is a periodic solution of the linear system

$$du/dt = A(t)u$$

By our choice of initial conditions for Y, we see that $p'(t)$ is the first column of $P(t)$ and L has the form shown.

The variation of constants formula shows that u satisfies the integral equation

$$u(t) = Y(t)u(0) + \int_0^t Y(t)Y^{-1}(s)g[p(s), u(s)]\,ds$$

We introduce one more change of variables: Let

$$w = P^{-1}(0)u$$

It follows that w satisfies the integral equation

$$w(t) = P^{-1}(0)Y(t)P(0)w(0) + \int_0^t P^{-1}(0)Y(t)Y^{-1}(s)g[p(s), P(0)w(s)]\,ds$$

We consider solutions of this equation that satisfy initial conditions of the form

$$w(0) = \text{col}(0, b_2, \cdots, b_N)$$

and let $\mathbf{b}$ be the vector ($\mathbf{b} \in E^{N-1}$) given by the last $N-1$ components of $w(0)$. The notation "col" indicates that $w(0)$ is a column vector having the given components.

With this, we have from Ref. 3.2 the following lemma.

Lemma. With the above notation and assumptions, for every vector $\mathbf{b}$ near zero, there is a unique number $T(\mathbf{b})$ such that the first component of w, say $w_1(t)$, satisfies

$$w_1[T(\mathbf{b})] = 0$$

and $T(\mathbf{b})$ is a continuous function of $\mathbf{b}$ for $\mathbf{b}$ near zero with $T(0) = T$, where T is the period of $p(t)$.

Proof. The solutions of the system near $p(t)$ are depicted in Figure 3.3:

The solutions beginning with $w_1(0) = 0$ lie on a plane that is orthogonal to $p(t)$ at the point $p(0)$. The fact that solutions depend continuously on initial data ensures that solutions starting with b near zero (i.e., starting near the orbit of p) remain near the orbit for $0 \le t \le 2T$. In particular, there are values of t near T, say t_1 and t_2 for which $w_1(t_1) < 0$ and $w_1(t_2) > 0$. The intermediate value theorem shows that there is a first value t, say t^*, such that t^* is near T and $w_1(t^*) = 0$. We define $T(\mathbf{b}) = t^*$. The continuity of $T(\mathbf{b})$ follows from the fact that the solutions of the system depend continuously on $\mathbf{b}$. ■

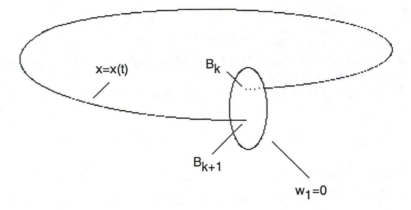

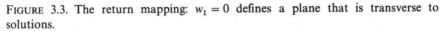

FIGURE 3.3. The return mapping: $w_1 = 0$ defines a plane that is transverse to solutions.

The Return Mapping

$T(\mathbf{b})$ is called the *return time* to the transversal plane defined by $w_1 = 0$. Let $\mathrm{col}(0, B_k)$ denote a point on this plane. Then $w[T(B_k)]$ defines a new point on the plane, say $\mathrm{col}[0, B_{k+1}]$. The mapping

$$B_k \to B_{k+1}$$

is called the *return mapping*. The fact that the return mapping is well-defined relies on the plane section being transversal to the flow defined by the solutions.

If the eigenvalues of the matrix Λ have negative real parts, then it follows that $B_k \to 0$ as $k \to \infty$. We summarize these results in the following Theorem:

Theorem. *Let $p(t)$ be a periodic solution of the equation*

$$dx/dt = f(x)$$

having least period $T > 0$ (in particular, p is not a rest point for the system). Suppose that f is a smooth function of the components of x. Moreover, suppose that $N - 1$ characteristic multipliers of the linear system

$$dx/dt = f_x[p(t)]x$$

have a modulus less than 1. Then $p(t)$ is orbitally asymptotically stable.

The proof of this result follows from the calculations presented above and the fact that if the vectors B_j approach zero, then the corresponding solutions of the system, $x(t, B_j)$, approach $p(t)$ uniformly on the interval $0 \le t \le T$. Important examples of how this result can be used are described later when we discuss the bifurcation of periodic solutions in Chapter 6.

Unfortunately, this result does not apply in some important cases. For example, consider the oscillator

$$x'' + f(x) = 0$$

where x is a scalar and f is a smooth function. Suppose that there is a 2π-periodic solution to this equation, say $x = p(t)$. Linearizing the equation about this oscillation leads to the equation

$$u'' + f'[p(t)]u = 0$$

which is a form of Hill's equation. In this case, we know that one solution of this equation is $u = p'(t)$, and by shifting the time variable we can take $p'(2\pi) = 1$. It follows that one of the characteristic multipliers of this equation has modulus 1, so the other one must also have modulus 1 (see Section 1.4.1). Thus, the conditions of the theorem are not satisfied.

3.6   Angular Phase Stability

The preceding discussion of amplitude stability should not hide the fact that phase stability is very important too. In fact, it is surprising that asymptotic stability of phases can occur in systems in which energy is conserved. This result is perhaps against intuition, but we illustrate it here by studying the clock problem formulated by van Huygens in the 1600s. But, we first describe a useful method for analyzing phase equations.

3.6.1   *Rotation Vector Method*

Let us consider a system of differential equations on an N torus

$$dx/dt = \omega + f(x)$$

where f is a continuously differentiable function that is periodic in each component of x, say with period 2π. We suppose that the mean value of f is zero, so that ω accounts for the mean value of the right-hand side of this system. Fourier series can be used to expand f:

$$f(x) = \sum_{|n| = -\infty}^{\infty} f_n e^{in \cdot x}$$

where the sum is taken over all multiindices $n \neq 0$. (Note that $f_0 = 0$.)

Let ω be nearly proportional to a vector of integers, say $\omega \sim \alpha \mathbf{n}$, where $\mathbf{n}$ is a vector of integers and α is the constant of proportionality.

We suppose first that there are positive constants ε and δ such that

$$\frac{|f(x)|}{|x|} \leq \varepsilon \quad \text{and} \quad \alpha^2 |\mathbf{n}|^2 + \mathbf{n} \cdot f(x) \geq \delta > 0 \tag{3.3}$$

for all x in T^N.

We define the components of the phase distribution vector to be

$$p_j(t) = \frac{x_j(t)/n_j}{\sum_{k=1}^{N} x_k(t)/n_k}$$

These ratios all approach $1/N$ as $t \to \infty$ if and only if the entropy function

$$H(p) = -\frac{1}{\log N} \sum_{j=1}^{N} p_j \log p_j$$

approaches a maximum value ($H = 1.0$). Equivalently,

$$x_1 : x_2 : \cdots : x_N \to n_1 : n_2 : \cdots : n_N$$

as $t \to \infty$. Thus, H, which is the entropy of the distribution p, provides a useful measure of synchronization of the oscillators.

Next, we define variables $v = \sum_{k=1}^{N} x_k(t)/n_k$ and

$$q_j(t) = v(Np_j(t) - 1)$$

A short calculation shows that

$$\frac{dq_j}{dv} = \varepsilon \left[G_j(\mathbf{q} + \mathbf{1}v) - (1/N) \sum_{k=1}^{N} G_k(\mathbf{q} + \mathbf{1}v) \right] + O(\varepsilon^2)$$

where $\mathbf{1}$ is the vector of ones. Since $\mathbf{q} \cdot \mathbf{1} = 0$, we can write $\mathbf{q} = Wu$, where W is an $(N-1) \times N$ matrix whose columns are $W_2, \cdots, W_N$, such that the set $\{\mathbf{1}, W_2, \cdots, W_N\}$ forms an orthogonal basis of E^N.

We return to this calculation in Chapter 7 where the method of averaging will show that a system of equations for the $N-1$ components of u results, namely

$$\frac{du}{dt} = \varepsilon F(Wu) + O(\varepsilon^2)$$

where F is periodic in the components of u.

We suppose that this system for u is stable under persistent disturbances. If this is the case, then for this and all nearby systems, u is bounded; therefore, so then is q. It follows that $p_j \to 1/N$ for each $j = 1, \cdots, N$.

The *rotation vector method* [3.15] comprises a collection of stability results about this system. The central idea is that if the functions q_j defined by these equations remain bounded as $v \to \infty$ and if $v \to \infty$ as $t \to \infty$, then the formula

$$q_j(t) = v(Np_j(t) - 1)$$

in the limit $t = \infty$ yields that the relative ratios of the components of x approach those of ω:

$$x_1 : \cdots : x_N \to \omega_1 : \cdots : \omega_N$$

as $t \to \infty$. The right-hand side of this limit is referred to as a *rotation vector*, in analogy with the rotation number of Denjoy [3.15, 3.16].

In later chapters, we derive useful conditions under which the system for u is stable under persistent disturbances. The result that

$$x_1 : \cdots : x_N \to \omega_1 : \cdots : \omega_N$$

as $t \to \infty$ implies that the system is *synchronized*.

If the system for u has a solution that is stable under persistent disturbances, then the same result holds for all nearby systems. This is the phenomenon of *phase-locking*; namely, the phases synchronize at the same relative frequencies even though the system is slightly changed. This is described in specific examples after we have discussed the method of averaging in Chapter 7.

3.6.2 Van Huygens's Problem

Let us consider a system of N pendula, where the angular deviation of the jth pendulum from rest is θ_j, which we suppose is not large. Then the system is described by the equations

$$\ddot{\theta}_j + \omega_j^2 \sin \theta_j = -\varepsilon F_j(\theta, \dot{\theta})$$

where F_j describes the force exerted on the jth pendulum by the system when it is in state $\theta, \dot{\theta}$. This problem has been studied in detail by Korteweg [3.17], and we consider here only an illustrative example. When $\varepsilon = 0$, there is no coupling, and the jth pendulum oscillates with frequency ω_j. This corresponds to each pendulum swinging in isolation from the others. When the pendula are coupled, they tend to synchronize. As an example, let us consider a lattice of oscillators having nearest neighbor interactions. To fix ideas, we take

$$F_j(\theta, \dot{\theta}) = \frac{\partial}{\partial \theta_j} \sum_{k=1}^{N} \cos(\theta_k - \theta_{k+1}) + I_j$$

where I_j is an applied torque and we take $\theta_{N+1} = \theta_1$.

In the case of $N = 3$, we have

$$\ddot{\theta}_1 + \omega_1^2 \sin \theta_1 = I_1 - \varepsilon[\sin(\theta_1 - \theta_2) - \sin(\theta_1 - \theta_3)]$$

$$\ddot{\theta}_2 + \omega_2^2 \sin \theta_2 = I_2 - \varepsilon[\sin(\theta_2 - \theta_3) - \sin(\theta_2 - \theta_1)]$$

$$\ddot{\theta}_3 + \omega_3^2 \sin \theta_3 = I_3 - \varepsilon[\sin(\theta_3 - \theta_1) - \sin(\theta_3 - \theta_2)]$$

We simulate this system using an interesting form of output: Select from a uniform distribution of values of the vector $\omega = (\omega_1, \omega_2, \omega_3)$ in the probability simplex, and for each calculate the corresponding solution. Then, normalize these vaues by setting $p_j = \theta_j/(\theta_1 + \theta_2 + \theta_3)$ and plot the results using areal coordinates.

Figure 3.4 shows the results of simulations of this system. In this case, we wish to consider the ratio $\theta_1 : \theta_2 : \theta_3$ as $t \to \infty$. These can be plotted simultaneously using triangular coordinates.

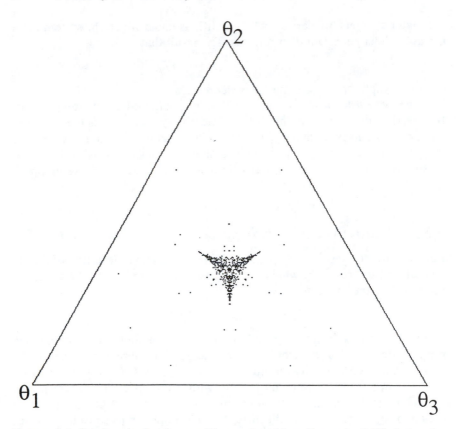

FIGURE 3.4. Synchronization for van Huygens' problem. For each of 400 choices for $(\omega_1, \omega_2, \omega_3)$, the problem for $(\theta_1, \theta_2, \theta_3)$ was solved using $\varepsilon = 0.5$ and $I_j = 0.1$ up to $t = 100\pi$. The ratios $p_j = \theta_j/(\theta_1 + \theta_2 + \theta_3)$ for $j = 1, 2, 3$, are plotted in a probability simplex. Most of the values are near where $p_1(t)\, p_2(t) : p_3(t) \sim 1:1:1$, which shows that the oscillators are near synchrony.

3.7   Exercises

3.1. a. Gronwall's inequality shows that if u, w, and v are continuous functions that are nonnegative and if

$$u(t) \leq v(t) + \int_0^t w(s)u(s)\,ds$$

then

$$u(t) \leq v(t) + \int_0^t w(s)v(s)\exp\left[\int_s^t w(s')\,ds'\right]ds$$

Prove this by defining $z(t) = \int_0^t w(s)u(s)\,ds$, showing that

$$dz/dt \leq wv + wz$$

and integrating this equation. Simplify the result in the case where $w =$ constant.

b. The following comparison theorem is quite useful: Let $x(t)$ solve the integral equation

$$x(t) = f(t) + \int_0^t k(t - s)x(s) \, ds$$

where f and k are smooth functions, and suppose that $z(t)$ satisfies the inequality

$$z(t) \leq |f(t)| + \int_0^t |k(t - s)| z(s) \, ds$$

and $z(0) = |x(0)|$. Show that $z(t) \geq |x(t)|$ for all t.

3.2. a. Show that a solution of Perron's integral equations in Section 3.1.5 defines a solution of differential equations in that section.

b*. Show that Perron's integral equations have a unique solution, and that it exists for all $0 \leq t < \infty$. Show that the solution of this system approaches zero as $t \to \infty$. (*Hint*: Use successive approximations and Gronwall's inequality.)

3.3. Suppose that $f(t, x)$ is a vector of N functions having period T in t and being continuously differentiable. Suppose that $x = 0$ is a solution of the differential equation

$$dx/dt = f(t, x)$$

That is, $f(t, 0) = 0$ for all t.

Show that if the solution of this equation is asymptotically stable, then it is uniform-asymptotically stable.

3.4. Consider the gradient system

$$dx/dt = -\text{grad } F(x).$$

Suppose that $x*$ is a local minimum for F. Show that $V(x) = F(x) - F(x*)$ defines a Liapunov function for this system near $x = x*$. Show that this gradient system cannot have a periodic solution. [*Hint*: Integrate $(dx/dt, dx/dt)(t)$ around an orbit.]

b. Apply the gradient method for finding the extrema of the function

$$F(x, y) = ax^2 + bxy + cy^2$$

in each of the three cases

i. $b = 0$ $ac > 0$ $a > 0$

ii. $a = c = 0$ $b > 0$

iii. $b = 0$ $ac < 0$

Discuss the general case.

c. Show that the gradient system with potential function

$$F(x) = -\sum x_k \log x_k$$

where the sum is taken for $k = 1, \cdots, N$, can be solved using the exponential integral function

$$Ei(u) = \int_{-\infty}^{u} \frac{e^v}{v}\, dv$$

Show that $x_j = (1/N)\exp[Ei^{-1}(t)]$, and so show that $x_j(t) \to 1/N$ as $t \to \infty$ for each $j = 1, \cdots, N$.

3.5. Find a matrix A having a positive eigenvalue while the eigenvalues of the symmetric part of A are negative. Use this example to show that $x^{tr}x$ might not define a Liapunov function for the system $dx/dt = Ax$.

3.6. Show that the return mapping in Section 3.5.4 has the form

$$B_{k+1} = \exp(\Lambda T)B_k + o(|B_k|).$$

Show that if all of the eigenvalues of Λ have negative real parts, then $|B_k| \to 0$ as $k \to \infty$. Show how the eigenvalues of Λ are related to the original system in Section 3.5.4.

3.7. Show that if $p(t)$ is a periodic solution of the system

$$dx/dt = f(x)$$

then the function $y = dp/dt$ defines a periodic solution of the linear problem

$$\frac{du}{dt} = \frac{\partial f}{\partial x}[p(t)]u$$

Conclude that at least one characteristic exponent of this system has real part zero, and so at least one characteristic multiplier has modulus equal to one.

3.8. Determine the (angular) phase stability of the system

$$\frac{dx}{dt} = \omega + \sin(x - y)$$

$$\frac{dy}{dt} = \mu + \sin(y - x)$$

In particular, determine for what values of ω and μ does $\rho = 1.0$.

4
Algebraic and Topological Aspects of Nonlinear Oscillations

It is often possible to reduce stability and oscillation problems for differential equations to problems for systems of algebraic equations. These can then be studied using implicit function theorems and fixed point arguments.

Consider a system of (algebraic) equations

$$F(u, \lambda) = 0$$

where $u \in E^N$ is the vector of unknowns, $\lambda \in E^K$ is a vector of parameters, and F is a vector of smooth functions, taking its values in E^N. Suppose that F has at least $M + 1$ continuous derivatives with respect to each of its arguments.

The problem studied in this chapter is to solve this system for u in terms of the parameters in λ. We first derive some implicit function theorems, and we then describe techniques for solving (nonlinear) bifurcation equations. These results are illustrated with examples of bifurcations of static and periodic solutions in systems of differential equations.

Finally we consider some geometric methods that are useful for finding solutions. Among these are some fixed point theorems from topology.

4.1   Implicit Function Theorems

Linear systems of equations are the easiest to solve, but even they are usually complicated. Solutions of them are first described, followed by nonlinear problems using implicit function methods that are based on solving linear problems.

4.1.1   *Fredholm's Alternative for Linear Problems*

Consider the equation for x

$$Ax = f$$

where $A \in E^{N \times N}$ and $f \in E^N$. The matrix A and the vector f are given, and we are to determine a vector x that solves this equation.

If the matrix A is invertible, then the solution is given by the formula

$$x = A^{-1}f$$

This formula can be evaluated using the spectral decomposition of A or by direct numerical evaluation, which may or may not be difficult. However, if the matrix A is not invertible, then other problems arise.

Suppose for the remainder of this section that A has rank $N - q$ and that there are q linearly independent vectors, say $\phi_1, \cdots, \phi_q$, that span the null space, or the *kernel* of A, which we denote by ker(A); that is, if v is a vector for which $Av = 0$, then there are unique constants $b_1, \cdots, b_q$ such that

$$v = b_1\phi_1 + \cdots + b_q\phi_q$$

We denote by ker(A)$^\perp$ the collection of all vectors that are orthogonal to ker(A); that is,

$$\ker(A)^\perp = \{v \in E^N : v \cdot \phi_j = 0 \text{ for } j = 1, \cdots, q\}$$

ker(A)$^\perp$ is called the *orthogonal complement* of ker(A), or the cokernel of A. From these definitions we have that

$$E^N = \ker(A) + \ker(A)^\perp$$

This means that any vector v in E^N can be written in a unique way as

$$v = c_1\phi_1 + \cdots + c_q\phi_q + w$$

where $w \in \ker(A)^\perp$ and $c_1, \cdots, c_q$ are uniquely defined constants.

We denote by A^{tr} the transpose of A, and by ker(A^{tr}) and ker(A^{tr})$^\perp$ the kernel of A^{tr} and its orthogonal complement, respectively.

Fredholm's Alternative. *This alternative is usually stated in the following way [see Ref. 4.1.]: Either the equation*

$$Ax = f$$

has a unique solution for any choice of the forcing vector f, or the homogeneous equation

$$Ax = 0$$

has a nonzero solution.

However, the alternative has come to mean more in the current literature. Figure 4.1 describes decompositions of E^N that A and A^{tr} define, and so it tells a great deal about solving for x.

Fredholm's Alternative comprises the following results, summarized in Figure 4.1.

1. The matrix A maps the set ker(A) into 0, since all of the vectors in ker(A) are null vectors.

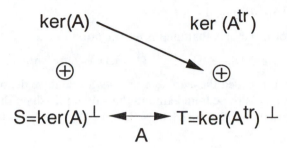

FIGURE 4.1. Diagram of the Fredholm Alternative.

2. A maps the set $S = \ker(A)^\perp$ one-to-one onto the set $T = \ker(A^{\text{tr}})^\perp$. That is, if $f \perp \ker(A^{\text{tr}})$, then there is a *unique* vector in $\ker(A)^\perp$, say w^*, such that $Aw^* = f$.
3. In order for there to be a solution to the equation $Ax = f$, f must lie in $\ker(A^{\text{tr}})^\perp$. In that case, the general solution has the form

$$x = c_1\phi_1 + \cdots + c_q\phi_q + w^*$$

where $c_1, \cdots, c_q$ are arbitrary constants, but w^* is uniquely determined by f.

Therefore, Fredholm's Alternative establishes a useful decomposition of E^N into the null space of A^{tr} and its complement. Since A does not map into any part of $\ker(A^{\text{tr}})$, the equation cannot be solved if part of f lies in this set. This can be seen simply by finding vectors $\psi_1, \cdots, \psi_q$ that span $\ker(A^{\text{tr}})$ and observing that if there is a solution (say, then $Ax = f$), then

$$\psi_j \cdot f = \psi_j \cdot Ax = A^{\text{tr}}\psi_j \cdot x = 0$$

for $j = 1, \cdots, q$. The equations

$$\psi_j \cdot f = 0$$

for $j = 1, \cdots, q$, are called the *solvability conditions* for this equation. It is necessary that f satisfy these conditions if there is to be a solution. In other words, they ensure that f is in $\ker(A^{\text{tr}})^\perp$.

Proof of Fredholm's Alternative.

Result 1 is obvious, and the first part of result 3 was just proved. It remains to show that A defines a one-to-one mapping of $S = \ker(A)^\perp$ onto $T = \ker(A^{\text{tr}})^\perp$. First, if w and w_1 are in S and if $Aw = Aw_1$, then $A(w - w_1) = 0$, so $w - w_1 \in \ker(A)$. Since S is a linear subspace of E^N and since 0 is the only element common to both $\ker(A)$ and S, it must be that $w - w_1 = 0$. Therefore, A defines a one-to-one mapping of S into T.

Now suppose that $f \in \ker(A^{\text{tr}})^\perp$. We must show that there is a vector $w^* \in S$ such that $Aw^* = f$. This can be done by actually constructing an approximate solution for general f.

Least-Squares Solution of $Ax = f$

We first ask for a vector x that minimizes the norm of $Ax - f$:

$$\eta = (Ax - f) \cdot (Ax - f) = |Ax - f|^2 = \min$$

If there is one, it is called the *least-squares solution*, and we denote it by x_{LSQ}. The value for x_{LSQ} is to be found among the values for which the gradient of this norm vanishes. Since $\eta = (A^{\mathrm{tr}} Ax - 2A^{\mathrm{tr}} f) \cdot x + f \cdot f$ we see that $\partial \eta / \partial x_j = 0$ for $j = 1, \cdots, N$, when

$$A^{\mathrm{tr}} Ax = A^{\mathrm{tr}} f$$

This system has certain advantages over the original one: The matrix on the left-hand side is now symmetric, so a spectral decomposition of $A^{\mathrm{tr}} A$ is available and it is useful. Also, the forcing vector $A^{\mathrm{tr}} f$ satisfies the solvability conditions.

The matrix $A^{\mathrm{tr}} A$ is a symmetric matrix of rank $N - q$ and nullity q. Therefore, there is an orthogonal matrix P ($P^{\mathrm{tr}} P = P P^{\mathrm{tr}} = I$) having the form

$$P = (\phi_1, \cdots, \phi_q, \xi_{q+1}, \cdots, \xi_N)$$

where the column vectors $\phi_1, \cdots, \phi_q$ span $\ker(A)$ such that P diagonalizes $A^{\mathrm{tr}} A$; that is

$$P^{\mathrm{tr}} A^{\mathrm{tr}} AP = \begin{pmatrix} 0 & 0 \\ 0 & D \end{pmatrix} \tag{4.1}$$

where D is a diagonal matrix, say $D = \mathrm{diag}(d_{q+1}, \cdots, d_N)$ [4.2]. $\det D \neq 0$, since the rank of D is $N - q$. Note that the columns of AP are $(0, \cdots, 0, A\xi_{q+1}, \cdots, A\xi_N)$. In fact, if we denote the ith column of the matrix AP by $(AP)_i$, then $(AP)_i = 0$ if $i = 1, \cdots, q$, but $(AP)_i = A\xi_i$ for $i = q + 1, \cdots, N$. Moreover, from Eq. (4.1) we see that

$$(AP)_i \cdot (AP)_j = (AP)_i^{\mathrm{tr}} (AP)_j = 0 \qquad \text{if } i \neq j \text{ or if } i = j \leq q$$

and

$$(AP)_j \cdot (AP)_j = d_j \qquad \text{if } j > q$$

The vectors $(AP)_{q+1}, \cdots, (AP)_N$ span $\ker(A^{\mathrm{tr}})^{\perp}$ since they are $N - q$ linearly independent elements of that subspace. Note that $(\xi_{q+1}, \cdots, \xi_N)$ are eigenvectors of A corresponding to nonzero eigenvalues.

Next, we define $z = P^{\mathrm{tr}} x$. With this, the equation becomes

$$P^{\mathrm{tr}} A^{\mathrm{tr}} APz = P^{\mathrm{tr}} A^{\mathrm{tr}} f$$

Hence,

$$\begin{pmatrix} 0 & 0 \\ 0 & D \end{pmatrix} z = (AP)^{\mathrm{tr}} f = (0, \cdots, 0, A\xi_{q+1}, \cdots, A\xi_N)^{\mathrm{tr}} f$$

Thus, the first q components of the right-hand side are zero, and the last are given by $(A\xi_j)^{\mathrm{tr}} f$ for $j = q + 1, \cdots, N$.

The first q equations show that $z_1, \cdots, z_q$ are free. The last $N - q$ equations show that $z_j = (A\xi_j) \cdot f/d_j$. Therefore,

$$x_{\text{LSQ}} = \sum_{i=1}^{q} z_i \phi_i + \sum_{i=q+1}^{N} (f \cdot A\xi_i) \xi_i / d_i$$

This vector solves the equation

$$A^{\text{tr}} A x_{\text{LSQ}} = A^{\text{tr}} f$$

Next, let

$$f_R = f - \sum_{j=1}^{q} C_j \psi_j$$

where $C_j = f \cdot \psi_j$. The vector f_R is the part of f that lies in $\ker(A^{\text{tr}})^\perp$. We have the following lemma.

Lemma. $A x_{\text{LSQ}} = f_R$.

Proof. The vectors $A\xi_{q+1}, \cdots, A\xi_N$, are orthogonal and they span $\ker(A^{\text{tr}})^\perp$. Therefore, f_R can be expanded in terms of them:

$$f_R = \sum_{i=q+1}^{N} f_i A\xi_i$$

where $f_j = f_R \cdot A\xi_j / d_j$. Since

$$x_{\text{LSQ}} = \sum_{i=1}^{q} z_i \phi_i + \sum_{i=q+1}^{N} (f \cdot A\xi_i) \xi_i / d_i$$

we have that

$$A x_{\text{LSQ}} = A \sum_{i=q+1}^{N} (f \cdot A\xi_i) \xi_i / d_i = \sum_{i=q+1}^{N} (f \cdot A\xi_i) A\xi_i / d_i = f_R \qquad \blacksquare$$

In summary, we see that the equation

$$Ax = f$$

has a solution if and only if $f \in \ker(A^{\text{tr}})^\perp$. In that case, the solution is given by the formula

$$x_{\text{LSQ}} = \sum_{i=1}^{q} z_i \phi_i + \sum_{i=q+1}^{N} (f \cdot A\xi_i) \xi_i / d_i$$

where $z_1, \cdots, z_q$ are free constants. In particular, if $f = f_R$, then there is a unique w^* in $\ker(A)^\perp$ such that $Aw^* = f$. This completes the proof of results 2 of Fredholm's Alternative, and the second part of 3. $\qquad \blacksquare$

It is tempting to use the least-squares method as the basis for a numerical scheme for finding x_{LSQ}. However, there are significant accuracy problems in this approach, and a more sophisticated approach is usually used [4.3].

4.1.2 *Nonlinear Problems: The Invertible Case*

Suppose that $u = u_0$ (in E^N) solves the equation

$$F(u_0, \lambda_0) = 0$$

for some choice of the parameters λ_0 in E^K. Without loss of generality, we can assume that $u_0 = 0$ and $\lambda_0 = 0$ since a simple translation of variables $(u \to u - u_0, \lambda \to \lambda - \lambda_0)$ gives a new system for which this is true. Thus, we suppose that

$$F(0, 0) = 0$$

Recall that the components of F are assumed to be at least $M + 1$ times continuously differentiable with respect to the components of u and of λ.

We look only for *small solutions* of this system for *small values of λ*; that is, we seek solutions that are near $u = 0$ for parameters near $\lambda = 0$. We begin by expanding the equation about this known solution:

$$F(u, \lambda) = C(\lambda) + A(\lambda)u + G(u, \lambda)$$

where $C(\lambda) = F(0, \lambda)$, $A(\lambda) = F_u(0, \lambda)$ is the Jacobian matrix, and $G(u, \lambda)$ is the remainder. There is a constant K so

$$|G(u, \lambda)| \le K|u|^2$$

for all small values of $|u|$ and $|\lambda|$. We consider here the case where $A(\lambda)$ is invertible, and, in the next section, the case where it is not.

Hypothesis HI. $A(0)$ is an invertible matrix.

With this we have the following result.

Implicit Function Theorem. *Suppose that condition HI and the above conditions are satisfied. Then there is a unique small solution, say $u = u(\lambda)$, of the equation*

$$F(u, \lambda) = 0$$

for λ near zero. In fact, there are functions $U_j(\lambda)$, homogeneous polynomials in the components of λ of degree j, for $j = 0, \cdots, M$, such that

$$u(\lambda) = \sum_{i=0}^{M} U_i(\lambda) + O(|\lambda|^{M+1})$$

The functions U_j have the form

$$U_j(\lambda) = \sum_{k_1 + \cdots + k_N = j} U(j, k_1, \cdots, k_N)\lambda_1^{k_1} \cdots \lambda_N^{k_N}$$

where the sum is taken over nonnegative indices $k_1, \cdots, k_K$ that satisfy the constraint.

Proof of the Implicit Function Theorem. The proof is accomplished in three steps.

Step 1: Determine the functions U_j recursively from the equation. Obviously, $U_0 = 0$ since we are looking for small solutions. Next, $U_1(\lambda) = A^{-1}G_\lambda(0,0)\lambda$, etc. These coefficients give Taylor's expansion of the expected solution.

Step 2: Let $v = u - \sum_{j=0}^M U_j(\lambda)$, and derive an equation for the remainder v. The result has the form

$$v = u - \sum_{i=0}^M U_i(\lambda) = -A^{-1}\left\{ G\left[v + \sum_{i=0}^M U_i(\lambda), \lambda \right] - G\left[\sum_{i=0}^M U_i(\lambda), \lambda \right] \right\}$$
$$+ O(|\lambda|^{M+1})$$

Thus, the remainder v solves the equation

$$v = H(v, \lambda)$$

where H is a smooth function with

$$H = O(|v|^2) + O(|\lambda||v|) + O(|\lambda|^{M+1})$$

as $|v| \to 0$ and $|\lambda| \to 0$.

Step 3: Finally, show that there is a unique solution for v and that $v = O(|\lambda|^{M+1})$ as $|\lambda| \to 0$. This is done by defining a sequence of successive approximations to v by $v_0 = 0$, and for $j = 1, 2, \cdots, v_j = H(v_{j-1}, \lambda)$. This defines a Cauchy sequence (see Section 4.4.1), and so it converges to a solution of the equation. To see that this solution satisfies the error estimate, we note that v_1 does, and that each further approximation does. In fact, we have that each v_j satisfies

$$|v_j| \le K|v_{j-1}|^2 + K|\lambda|^{M+1}$$

For some constant K, all small λ, and v_{j-1} satisfying $|\lambda| + |v_{j-1}| \le \delta$. Therefore,

$$|v_1| \le K|\lambda|^{M+1}$$
$$|v_2| \le K|v_1|^2 + K|\lambda|^{M+1} \le K[K^2|\lambda|^{2(M+1)} + K|\lambda|^{M+1}]$$

and by a straightforward induction

$$|v_j| \le K[r + r^2 + \cdots + r^j]$$

where $r = K|\lambda|^{M+1}$. It follows that v satisfies the estimate

$$|v| \le \frac{Kr}{1-r} = O(|\lambda|^{M+1}) \qquad \blacksquare$$

4.1.3   *Nonlinear Problems: The Noninvertible Case*

Now suppose that the linear part of the problem is not invertible. Specifically, suppose the following condition

Hypothesis HNI: $A(0)x = 0$ has q linearly independent solutions for x.

With this, $\ker[A(0)]$ is spanned by some vectors, $\phi_1, \cdots, \phi_q$, and $\ker[A(0)^{\mathrm{tr}}]$ is spanned by $\psi_1, \cdots, \psi_q$. We suppose that these are biorthogonal sets with

$$\psi_j \cdot \phi_k = \delta_{j,k}$$

for $j, k = 1, \cdots, q$, where $\delta_{j,k}$ is Kronecker's delta function. ($\delta_{j,k} = 1$ if $k = j$, $= 0$ otherwise.)

As in Section 4.1, we let

$$u = c_1 \phi_1 + \cdots + c_q \phi_q + w$$

where $c_1, \cdots, c_q$, are arbitrary constants and w is a vector that is orthogonal to the null space of $A(0)$: $w \cdot \phi_j = 0$ for $j = 1, \cdots, q$.

The equations to be solved become

$$A(0)w = H(c_1, \cdots, c_q, w, \lambda)$$
$$\equiv -C(\lambda) - [A(\lambda) - A(0)]u - G(c_1 \phi_1 + \cdots + c_q \phi_q + w, \lambda)$$

Here and below we write A for $A(0)$, and we use the matrix notation $\Phi c = c_1 \phi_1 + \cdots + c_q \phi_q$, where Φ is an $N \times q$ matrix whose columns are the vectors $\phi_1, \cdots, \phi_q$ and c is the column vector whose components are $c_1, \cdots, c_q$.

The invertible form of the implicit function theorem can be used to prove the existence of a unique function $w^*(c_1, \cdots, c_q, \lambda)$ that solves the equation

$$A^{\mathrm{tr}} Aw = -A^{\mathrm{tr}} H(c_1, \cdots, c_q, w, \lambda)$$

and is orthogonal to the null space of A. This is the least-squares version of the original equations. The solution w^* is a solution of the original system if the solvability conditions

$$\psi_j \cdot H[c_1, \cdots, c_q, w^*(c, \lambda), \lambda] = 0$$

are satisfied for $j = 1, \cdots, q$. This fact is a restatement of Fredholm's Alternative. These q equations for the q unknowns $c_1, \cdots, c_q$, are called the *bifurcation equations*. Each small solution of them for small λ gives a small solution of the original problem: namely,

$$u = \Phi c + w^*(c, \lambda)$$

This approach has been extended to study quite sophisticated problems in functional analysis, and it is now referred to as the *Liapunov-Schmidt Method* [see 4.4].

Figure 4.1 is important to keep in mind since it describes where w^* and the bifurcation equations come from. The combinations of $c_1, \cdots, c_q$, are selected by the solvability conditions to ensure that H lies in $T = \ker(A^T)^\perp$. This may or may not be possible. If so, then the invertible form of the implicit function theorem is used in the portion of E^N where A is invertible to find $w^*(c, \lambda)$ in $S = \ker(A)^\perp$. It is straightforward to derive w^* in practice by iteration, but solving the bifurcation equations is usually difficult.

4.2 Solving Some Bifurcation Equations

Finding u in the invertible case and w^* in the noninvertible case are usually straightforward, and the method of iteration described in the proof gives a reliable iterative scheme for finding them. However, in the noninvertible case, the bifurcation equations must be solved as well, and it is at this point that the hard work begins. There are no known general methods for doing this, but Newton's polygons are quite useful for a variety of problems.

4.2.1 $q = 1$: *Newton's Polygons*

When $q = 1$, the solution takes the form

$$x = c\phi + w(c, \lambda)$$

and the bifurcation equation becomes the scalar equation

$$0 = \psi \cdot f[c\phi + w(c, \lambda), \lambda]$$

which can be expanded in powers of c and λ. Therefore, we consider the scalar equation

$$0 = a_{0,0} + a_{1,0}c + a_{0,1}\lambda + a_{2,0}c^2 + a_{1,1}c\lambda + a_{0,2}\lambda^2 + \cdots$$

which we want to solve for *small real solutions* c when λ *is near* 0. Here are some examples:

1. There may be a unique small solution

$$0 = c + \lambda c^2 + \lambda^2 c + \lambda c$$

$$= c(1 + \lambda + \lambda^2 + c)$$

   Here $c = 0$ is the only small solution for λ near zero.
2. There may be more than one solution, and both could be an analytic function of λ:

$$0 = \lambda c^2 + \lambda^2 c$$

$$= \lambda c(c + \lambda)$$

   Thus, $c = 0$ and $c = -\lambda$ are two small solutions.
3. There may be several solutions, but they are analytic in some (fractional) power of λ:

$$0 = \lambda c^3 - \lambda^2 c$$

$$= \lambda c(c^2 - \lambda)$$

   Here $c = 0$, $c = \sqrt{\lambda}$ and $c = -\sqrt{\lambda}$ are small solutions.

   These examples suggest that we look for solutions in the form

$$c = \alpha \lambda^r + \text{higher order terms in } \lambda$$

for amplitude α and some base exponent r. Thus,

$$0 = a_{1,0}\alpha\lambda^r + a_{0,1}\lambda + a_{2,0}\alpha^2\lambda^{r2} + a_{1,1}\alpha\lambda^r\lambda + a_{0,2}\lambda^2 + \cdots$$
$$= (\cdots)\lambda^{r*} + \cdots$$

where

$$r* = \min\{rj + k = \text{constant}: j, k = 0, 1, 2, \cdots, \text{for which } a_{j,k} \neq 0\}$$

Terms that will balance to contribute to the coefficient of λ^{r*} can be determined by finding the indices (j, k) of nonzero coefficients for which at least two lie on the same straight line: $k = -rj + \text{constant}$. This can be done graphically by plotting k vertically and j horizontally, for which $a_{j,k} \neq 0$. Then the straight lines bounding this set of (j, k) values from $(0, 0)$ are determined and the slopes of these lines give the possible values of $-r$. Only lines having at least two points are of interest to get possible equations for α.

For example, consider the equation

$$0 = \lambda^2 + \lambda c - c^3 + c^4 + \lambda^2 c^2$$

Figure 4.2 shows the indices and the enclosing straight lines.

From Figure 4.2, we see that the two possible cases are

$$c \sim \lambda \qquad \text{and} \qquad c \sim \sqrt{\lambda}$$

Newton's polygon method rests on the following theorem, which shows that corresponding to each of these scalings, there is an analytic solution.

Weierstrass Preparation Theorem. *Suppose that $G(c, \lambda)$ is an analytic function of c and λ near $c = 0$, $\lambda = 0$, such that $G(0, 0) = 0$ and*

$$\frac{\partial^j G}{\partial c^j}(0, 0) = 0 \qquad \text{for } j = 1, \cdots, k - 1, \qquad \text{but } \frac{\partial^k G}{\partial c^k}(0, 0) \neq 0.$$

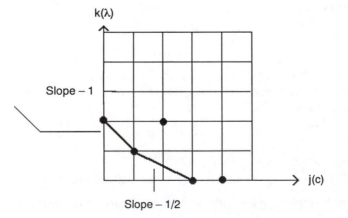

FIGURE 4.2. Newton's polygon for $\lambda^2 + \lambda c - c^3 + c^4 + \lambda^2 c^2$. There are two relevant slopes.

Then there are analytic functions $A_0(\lambda), \cdots, A_{k-1}(\lambda)$, and $B(c, \lambda)$ with $B(c, \lambda) \neq$
0 for c and λ near zero, such that

$$G(c, \lambda) = [c^k + A_{k-1}(\lambda)c^{k-1} + \cdots + A_0(\lambda)]B(c, \lambda)$$

We do not prove this result here [see 4.5]. If we wish to solve the equation

$$G(c, \lambda) = 0$$

for c as a function of λ, then since $B \neq 0$, the roots are determined by solving
the polynomial

$$c^k + A_{k-1}(\lambda)c^{k-1} + \cdots + A_0(\lambda) = 0$$

There are exactly k solutions of this equation for each value of λ near $\lambda = 0$.
Moreover, each root is an analytic function of some (fractional) power of λ.

If in the example depicted in Figure 4.2 we set

$$c = \alpha\sqrt{\lambda} + \beta\lambda + \cdots$$

then

$$\alpha - \alpha^3 = 0$$

$$(1 - 3\alpha^2)\beta + (\alpha^2 + \alpha^4) = 0$$

and so on. This shows that $\alpha = \pm 1$, 0, are the possible candidates for
starting; thereafter, all coefficients are uniquely determined. We will apply
this method in the next examples.

4.3   Examples of Bifurcation Equations

The two examples presented in this section illustrate two important bifurcation phenomena that arise in many applications. The first is a system that
exhibits the classical cusp bifurcation of static states, and the second illustrates a Hopf bifurcation of a periodic solution.

4.3.1   *Exchange of Stabilities*

Consider the system of differential equations

$$dx/dt = \lambda x - ayx$$

$$dy/dt = -by + dyx + fx^2$$

where a, b, d, and f are positive constants. Small static states of these equations are to be found by solving the equations $dx/dt = 0$, $dy/dt = 0$ for x and
y when λ is near zero. In this case, the matrix $A(\lambda) = \text{diag}(\lambda, -b)$, and when
$\lambda = 0$, $\phi = \psi = (1, 0)^{\text{tr}}$. Therefore, we seek the solutions of the form $x = c$,
$y = w$, in the notation of Section 4.1.3.

The first step in using the Liapunov-Schmidt method involves solving

$$0 = -bw + dwc + fc^2 + \text{h.o.t.}$$

for w as a function of c and λ, where h.o.t. denotes higher order terms in c. Obviously,

$$w = fc^2/b + \text{h.o.t.}$$

Substituting this result into the first equation gives the bifurcation equation

$$0 = \lambda c - (af/b)c^3 + c(\text{h.o.t.})$$

Using Newton's polygons, we see that, if $\lambda < 0$, then $c = 0$ is the only (small) real solution. If $\lambda > 0$, then there are three possible solutions, $c = 0$ and $c = \pm\sqrt{\frac{\lambda b}{af}} + \text{h.o.t.}$

Therefore, this system has a unique small static state for $\lambda < 0$, namely $x = y = 0$, and there are three small static states for $\lambda > 0$. An interesting aspect of this example is that the solution $x = y = 0$ is exponentially stable for $\lambda < 0$, but it is unstable for $\lambda > 0$. However, for $\lambda > 0$, the two new (nonzero) solutions are both exponentially stable. This behavior is typical of many stability problems, and it is referred to as being an *exchange of stabilities*. Note that when $\lambda = 0$, the solution $x = y = 0$ is UAS but not EAS. This is studied using other methods in Chapter 8 [see 4.6, 4.7].

4.3.2   A $\lambda\omega$ System

The following example is quite useful.

$$dx/dt = [\lambda - (x^2 + y^2)]x - \omega y$$
$$dy/dt = \omega x + [\lambda - (x^2 + y^2)]y$$

where λ and ω are constants. Setting $r^2 = x^2 + y^2$ and $\theta = \tan^{-1}(y/x)$, this problem becomes

$$dr/dt = r(\lambda - r^2)$$
$$d\theta/dt = \omega$$

Dividing the first equation by the second gives

$$\frac{dr}{d\theta} = \frac{r(\lambda - r^2)}{\omega}$$

Thus, the $\lambda\omega$ system reduces easily to a scalar differential equation. Static states for this equation ($r = r^*$) correspond to periodic solutions of the original system. We considered the same static state problem in the preceding section. For $\lambda < 0$, there is a unique real static state for r, namely $r = 0$. For $\lambda > 0$ there are three real solutions $r = 0$ and $r = \pm\sqrt{\lambda}$. The two nonzero static states for r correspond to periodic solutions, one π out of phase with the other, so we consider only the positive root. The result for the original system is that for $\lambda > 0$ there is a unique periodic orbit, namely

$$x = \sqrt{\lambda}\cos(\omega t + \phi) \qquad y = \sqrt{\lambda}\sin(\omega t + \phi)$$

where ϕ is a free constant.

We say that as λ increases through $\lambda = 0$, a periodic solution bifurcates from the static state $x = 0$, $y = 0$. Note that this static state is stable for $\lambda < 0$ but unstable for $\lambda > 0$; however, the periodic solution is orbitally exponentially stable for $\lambda > 0$. Therefore, an exchange of stabilities can also occur through the appearance of a periodic solution.

The $\lambda\omega$ system provides a good example of a periodic solution bifurcation, and it is typical of a class of problems called Hopf bifurcation problems. We study these later in Chapter 8 [see 4.8].

4.4 Fixed-Point Theorems

A forced system has the form

$$dx/dt = f(t, x)$$

where x and f are in E^N, and we suppose that f is a smooth function of its arguments having (least) period T in t.

A solution starting at a point $x(0)$ evolves into a point $x(T)$ after T units, and as in Chapter 3, we define the *return mapping*, or Poincare mapping, by

$$P: x(0) \to x(T)$$

Fixed points of this mapping correspond to periodic solutions of the original system of equations, and a study of P helps us to clarify several important issues. The following lemma shows this.

Lemma. P has a fixed point if and only if the differential equation used to define it has a solution of period T (not necessarily its least period).

Proof. If $Py = y$, then let the solution of the defining system with initial value $x(0) = y$ be denoted $x(t)$. Let $z(t) = x(t + T)$. Note that $z(t)$ satisfies the same differential equation as x, and that $z(0) = x(T) = y$. Since solutions of this problem are unique, x and z are identical. This shows that $x(t) = x(t + T)$ for all t. Conversely, if $x(t)$ is a solution having period T, then $x(0)$ is a fixed point of P. ∎

This lemma shows that a fixed point of the return mapping corresponds to a periodic solution of the original system. Therefore, we can establish the existence of periodic solutions to the system of differential equations by finding the fixed points of this mapping. Several methods are available for doing this.

4.4.1 *Contraction Mapping Principle*

Suppose that F is a continuous mapping of a closed subset Ω of E^N into itself such that for some constant λ, $0 \le \lambda < 1$, we have that

$$|F(x) - F(y)| \le \lambda |x - y|$$

for all x and y in Ω. Then there is a unique fixed point for F in Ω, and it is asymptotically stable under iterations of F.

The proof of this result begins with showing that the iteration sequence $\{x_n\}$ defined by

$$x_{n+1} = F(x_n)$$

is a Cauchy sequence. In fact, $|x_{n+1} - x_n| \le \lambda |x_n - x_{n-1}|$, so $|x_{n+1} - x_n| \le \lambda^n |x_1 - x_0|$. It follows that for any integers k and n,

$$|x_{n+k} - x_n| \le \left(\sum_{j=n}^{n+k} \lambda^j \right) |x_1 - x_0|$$

Since the sum in this estimate approaches zero as n and $k \to \infty$, we see that $\{x_n\}$ is a Cauchy sequence. Therefore, it converges to a limit, say x^*. Passing to the limit $n \to \infty$ in the formula

$$x_{n+1} = F(x_n)$$

shows that $x^* = F(x^*)$. Obviously, x^* is the only fixed point of F. Finally, an initial condition, say y_0, lying near x_0, defines an iteration sequence $\{y_n\}$. Since

$$|y_n - x_n| \le \lambda^n |y_0 - x_0|,$$

we see that $y_n \to x^*$ as $n \to \infty$. Thus, the equilibrium is asymptotically stable [see 4.9].

Example. A Forced Stable Linear System. The equation

$$dx/dt = Ax + g(t)$$

where A is a constant matrix having all eigenvalues in the left half-plane and g is a smooth function having period T, has a unique periodic solution. This can be proved using the contraction mapping principle, as we show next.

The return mapping is defined for this system as follows: An initial value $x(0)$ determines an unique solution of this equation; namely,

$$x(t) = \exp(At)x(0) + \int_0^t \exp[A(t - s)]g(s)\,ds$$

The return mapping therefore is

$$P(\xi) = \exp(AT)\xi + \int_0^T \exp[A(T - s)]g(s)\,ds$$

where we write $x(0) = \xi$. Since A is a stable matrix, we have

$$|\exp(ATm)| < 1$$

for sufficiently large values of m. Therefore, P^m is a contraction mapping for sufficiently large m. It follows that P^m has a unique fixed point, say ξ^*. Also, ξ^* is a unique fixed point of P, since for any point $x(0)$ we have

$$P^{mk}[x(0)] \to \xi^*$$

as $k \to \infty$. Therefore,

$$P(\xi^*) = P[P^{mk}(\xi^*)] = P^{mk}[P(\xi^*)] \to \xi^*$$

as $k \to \infty$. It must be that

$$\xi^* = [I - \exp(AT)]^{-1} \int_0^T \exp[A(T - s)]g(s)\,ds$$

It follows that the differential equation has a unique periodic solution; namely,

$$x(t) = \exp(At)\xi^* + \int_0^t \exp[A(t - s)]g(s)\,ds$$

4.4.2   *Wazewski's Method*

A mapping need not be a contraction to have a fixed point. In fact, recall from calculus that every continuous function attains all values between its maximum and minimum values. If f is a continuous function mapping a finite interval $[a, b]$ into itself, then the graph of f (i.e., the set $\{[x, f(x)]: x \in [a, b]\}$) must intersect the one-to-one line $\{(x, x)\}$. Such an intersection defines a fixed point of f.

This observation is based on the intermediate value theorem of calculus, and it is possible to extend it to dimensions higher than one. This is done in Brouwer's fixed-point theorem.

Brouwer's Fixed-Point Theorem. *Suppose that F is a continuous mapping of a closed ball, say $\Omega \subset E^N$, into itself. Then F has a fixed point, say x^*, in Ω.*

Note that the fixed point given by this result need not be unique since the conditions are satisfied by the identity mapping on Ω. This result is not proved here [see 4.9].

Example. Forcing a system that is stable under persistent disturbances. Consider the equation

$$dx/dt = -x^3 + A\cos t$$

In the (t, x)-plane, we see that $x' < 0$ on the line $x = 2A$, but on the line

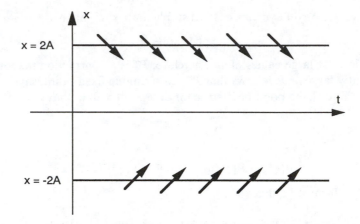

FIGURE 4.3. Example of Brouwer's theorem.

$x = -2A$, $x' > 0$. Therefore, solutions starting in the interval $-2A \leq x \leq 2A$ remain there for all future times. In particular, this interval is mapped into itself by the return mapping. It follows from Brouwer's theorem that the return mapping has a fixed point, and so there is a periodic solution of this equation.

A corollary of Brouwer's theorem is the No-Retract Theorem.

No-Retract Theorem. *There is no continuous mapping of the unit ball (or any compact manifold with boundary) in E^N onto its boundary that leaves boundary points fixed.*

Proof. If there were such a mapping, say G, of the set

$$B_N = \{x \in E^N : |x| \leq 1\}$$

called the unit ball in E^N, onto its boundary

$$S^{N-1} = \{x \in E^N : |x| = 1\}$$

then G followed by the antipodal mapping $x \to -x$ defines a continuous mapping of the unit ball into itself that has no fixed points. This contradicts Brouwer's Theorem, and so the No-Retract Theorem is proved.   ∎

Another interesting consequence of this theorem is useful in studying nonlinear differential equations. Consider solutions of a system of N equations

$$dx/dt = f(t, x)$$

that have initial values in a smooth, bounded, and simply connected domain Ω. Let c denote the cylinder in (t, x) space that is defined by

$$C = \{(t, x): 0 \leq t < \infty, x \text{ in } \Omega\}$$

Let ∂C denote the lateral boundary of this cylinder:

$$\partial C = \{(t, x): 0 \le t < \infty, x \text{ in } \partial\Omega\}$$

An *egress point* of C is a point in ∂C at which a solution of the system leaves the interior of C. A *strict egress point*, say (t^*, x^*), is one where a solution of the system lies inside C for $t^* - \delta < t < t^*$ and outside C for $t^* < t < t^* + \delta$ for some small positive number δ. With these definitions, we can discuss Wazewski's result:

Wazewski's Theorem. *Suppose that all egress points of the system are strict egress points. Then there is a solution of the system that begins in Ω and remains in C for all $t \ge 0$.*

Proof. If such a point did not exist, then every point in Ω defines an orbit that eventually leaves C. We define a mapping of Ω into ∂C by the solutions: If $p \in \Omega$ and $x(t, p)$ is the solution emanating from that point at $t = 0$, then the mapping is given by $p \to q = x(t^*, p)$ where t^* is the first time this solution hits ∂C. This is a continuous mapping since egress points are strict. The mapping $p \to q$ followed by the projection

$$(t^*, x^*) \to (0, x^*)$$

which is called a *pull-back mapping*, then defines a continuous mapping of Ω onto its boundary that leaves boundary points fixed. This contradicts the No-Retract Theorem and completes the proof of the theorem [see 4.9]. ∎

The solutions described by this theorem are usually difficult to construct numerically because they are often not stable. Some techniques based on a related theorem, Sperner's lemma, have been devised [see 4.10 and section 4.4.3].

Example. A Forced Unstable System. Consider the equation

$$dx/dt = x + A \cos t$$

The cylinder $\{(t, x): t \ge 0, -2A \le x \le 2A\}$ is bounded (in x) by the lines $x = \pm 2A$. Since every egress point is a strict egress point for this cylinder, there is a solution, say $x = p(t)$, that remains within these limits for $t \ge 0$.

This must be a periodic solution. Indeed, the general solution is given by the formula

$$x(t) = \exp(t)x(0) + \int_0^t \exp(t - s)A \cos s \, ds$$

The return mapping is given by

$$P(\xi) = \exp(2\pi)\xi + \int_0^{2\pi} \exp(2\pi - s)A \cos s \, ds$$

and it has a unique fixed point

$$\xi^* = [1 - \exp(2\pi)]^{-1} \int_0^{2\pi} \exp(2\pi - s) A \cos s \, ds$$

$p(t)$ is the solution that has the initial value ξ^*.

4.4.3   Sperner's Method

Suppose that f is a mapping of the plane E^2 into itself. Thus, for each point $x \in E^2$, f defines a vector

$$f(x) = [f_1(x_1, x_2), f_2(x_1, x_2)]$$

Consider a triangle T in E^2, select one side of T, say S, and trace the vector f around T keeping track of the angle it makes with the side S. The total angle passed through must be $2\pi N$ for some integer N. This is called the *index* of T relative to f. We denote it by $I(f, T)$. If T is a small triangle, we see that, if $f \neq 0$ in T and on its boundary, then $I(f, T) = 0$.

Next, consider two triangles, say T_1 and T_2 that share a side, say S, and suppose that $I(f, T_1) = 0$ and $I(f, T_2) = 0$. Let the part of the index of f that occurs on $T_2 - S$ be δ_2, then the part on S is $-\delta_2$. Similarly, if the index on $T_1 - S$ is δ_1, then the index on S is $-\delta_1$. It follows that

$$I(f, T_1 \cup T_2) = \delta_1 + \delta_2 = 0$$

Finally, consider an arbitrary triangle T. Suppose that $I(f, T) \neq 0$. Then there must be a point $p \in T$ such that $f(p) = 0$. Such a value can be found by introducing a triangulation of T into subtriangles of size d (i.e., the longest side of one of the partitioning triangles is length d). The result of the above argument is that for at least one of the triangles in this partition, say s, we must have $I(f, s) \neq 0$.

The index and this triangulation construction gives a method for finding a fixed point of f in a triangle T. First, select a mesh size d and a corresponding triangulation of T. Begin on the boundary of T. Successively testing subtriangles around the boundary for their index. Throw away triangles **t** for which $I(f, \mathbf{t}) = 0$. This algorithm, referred to as Sperner's method, leads us to a subtriangle, say s, for which $I(f, s) \neq 0$. This algorithm can be quite useful for computations [4.11, 4.12].

4.4.4   Measure-Preserving Mappings

Consider an externally forced Hamiltonian system

$$dx/dt = J\nabla H(x) + g(t)$$

where $x \in E^{2N}$, say $x = \text{col}(p, q)$ where each of $p, q \in E^N$, J is Jacobi's matrix

$$J = \begin{pmatrix} 0 & -I \\ I & 0 \end{pmatrix}$$

and H and g are smooth functions. Since the divergence of the right-hand side of this system is zero, the flow defines a volume preserving transformation of E^{2N} into itself (see Section 2.3.8). Suppose also that g has period T. Then the return mapping is defined by

$$x(0) \to x(T)$$

and it defines a measure-preserving transformation of E^{2N} into itself. We have seen in this chapter that fixed points of this mapping define periodic solutions of the original system. We have seen in Chapter 2, for example from Poincaré's twist theorem in Section 2.5, that the fixed-point structure of volume-preserving mappings can be quite complicated. This problem is the object of intense study.

4.5   Exercises

4.1. a. Show that the rank of an $N \times N$ matrix A and its transpose A^{tr} are the same. Show that if the equation $Ax = f$ is solvable, then f must be orthogonal to all null vectors of A^{tr}.

   b. Let $L = (f - Ax) \cdot (f - Ax)$. Show that the equations $\partial L / \partial x_j = 0$ for $j = 1, \cdots , N$, are equivalent to the equation

$$A^{tr} Ax = A^{tr} f$$

   c. Solve the system

$$2x + y = f_1$$
$$4x + 2y = f_2$$

   using the least-squares method described in part b.

   d*. Show that the least squares method works as well when A is in $E^{m \times n}$ where $m \neq n$.

4.2. Show that the sequence of iterates defined by the iteration in Section 4.1.2, say $\{v_j\}$, defines a Cauchy sequence, and so that it converges. Verify that the limit satisfies the estimate

$$v = O(|\lambda|^{M+1}) \qquad \text{as } \lambda \to 0$$

4.3. a. Prove the Weierstrass Preparation Theorem by showing that the function G defined in Section 4.2.1 can be written in the form

$$G(c, \lambda) = (c^k + \text{l.o.t.}) B(c, \lambda)$$

   where $B(0,0) \neq 0$. Here l.o.t. denotes lower order terms in powers of c.

   b. Construct all small solutions for c and λ near zero of the equation

$$c^9 - c^4 \lambda + c\lambda^2 = 0$$

   using Newton's polygons and the Weierstrass Preparation Theorem.

4.4. Exchange of stabilities. Verify the computations done in Section 4.3.1 and 4.3.2.

4.5. Consider the differential-delay equation

$$\frac{dx}{dt} = \lambda x(t-1)[1 - x^2(t)]$$

show that as λ increases through the value $\lambda = \pi/2$, a bifurcation of a periodic solution occurs. Do this by showing that as λ increases through the value $\lambda = \pi/2$ a single pair of eigenvalues of the linear problem

$$dx/dt = \lambda x(t-1)$$

crosses the imaginary axis from left to right. Show that for $\lambda = \pi/2$, the function $x = \sin \pi t/2$ solves the linear problem.

4.6. Verify that the formulas for ξ^* in the examples in Sections 4.4.1 and 4.4.2 lead to the definition of solutions having period T.

4.7. Verify that the graph of a continuous function f that maps an interval into itself must cross the one-to-one line at some point. Thus, the function must have a fixed point.

4.8. a. Solve the equation

$$\frac{dx}{dt} = -x^3 + A \cos t$$

numerically and plot $x(2\pi)$ versus $x(0)$. Show that there is a unique periodic solution of this equation and that it is asymptotically stable.

b. Show that the strict egress point mapping in Wazewski's theorem is a continuous mapping of the initial set into the boundary.

c. Use Wazewski's theorem to show that the differential equation

$$\frac{dx}{dt} = x^3 + A \cos t$$

has a unique periodic solution that starts and remains in the interval $-2A \le x(t) \le 2A$.

II
Perturbation Methods

5
Regular Perturbation Methods

Four kinds of perturbation problems are studied in the remainder of this book. They are illustrated by the following examples:

1. *Smooth Data and a Finite Interval.* The initial value problem

$$\frac{dx}{dt} = -x^3 + \varepsilon \qquad x(0, \varepsilon) = \xi_0 + \varepsilon\xi_1 + O(\varepsilon^2)$$

where the right-hand side defines a smooth function of ε at $\varepsilon = 0$, arises frequently in bifurcation problems. A solution can be constructed using Taylor's formula. First, when $\varepsilon = 0$, the solution is

$$x_0(t) = \frac{\xi_0}{\sqrt{1 + 2t\xi_0^2}}$$

and for $\varepsilon \neq 0$ it has the form

$$x(t) = x_0(t) + \varepsilon x_1(t) + O(\varepsilon^2)$$

where $x_1(t)$ solves the problem

$$\frac{dx_1}{dt} = -3x_0(t)x_1 + 1 \qquad x_1(0) = \xi_1$$

and so on. This is a *regular perturbation problem on a finite interval* $[0, T]$, and the error estimate $O(\varepsilon^2)$ holds as $\varepsilon \to 0$ uniformly for $0 \leq t \leq T$. However, since the solution

$$x(t) \to \sqrt{\varepsilon} \qquad \text{as } t \to \infty$$

we cannot expect Taylor's formula to be valid uniformly for $0 \leq t < \infty$.

2. *Smooth Data and an Infinite Interval.* The problem

$$\frac{dx}{dt} = -x + \varepsilon \qquad x(0, \varepsilon) = \xi_0 + \varepsilon\xi_1 + O(\varepsilon^2)$$

has the property that when $\varepsilon = 0$ the solution $x = 0$ is exponentially asymptotically stable. The solution of this equation is

$$x(t) = e^{-t}[\xi(\varepsilon) - \varepsilon] + \varepsilon$$

Since this depends smoothly on ε uniformly for $0 \leq t < \infty$, the problem is referred to as being a *regular perturbation problem on* $[0, \infty)$.

3. *Highly Oscillatory Data.* The coefficient of x in the problem

$$\frac{dx}{dt} = -\cos\left(\frac{t}{\varepsilon}\right)x + \varepsilon \qquad x(0, \varepsilon) = \xi_0 + \varepsilon\xi_1 + O(\varepsilon^2)$$

is highly oscillatory having period $2\pi\varepsilon$. This coefficient has an essential singularity at $\varepsilon = 0$, and the solution is

$$x(t) = \exp\left[\int_0^t \cos\left(\frac{s}{\varepsilon}\right) ds\right]x(0, \varepsilon) + \varepsilon \int_0^t \exp\left[\int_s^t \cos\left(\frac{s'}{\varepsilon}\right) ds'\right] ds$$

It is not apparent at first glance how one might extract useful information from this explicit formula. However, we will see that the Method of Averaging (described in Chapter 7) provides a method for doing this.

4. *Smooth Data—Singular Solutions.* Solutions might decay or grow rapidly over short time intervals. For example, the equation

$$\varepsilon\frac{dx}{dt} = -x \qquad x(0, \varepsilon) = \xi_0 + \varepsilon\xi_1 + O(\varepsilon^2)$$

appears to involve a smooth perturbation, but its solution is

$$x(t) = \exp(-t/\varepsilon)x(0, \varepsilon)$$

which has an essential singularity at $\varepsilon = 0$. Problems of this kind are studied in Chapter 8 using quasistatic state approximation methods.

Examples 1 and 2 are *regular perturbation problems* and 3 and 4 are *singular perturbation problems*, so called because their solutions depend regularly or singularly on ε at $\varepsilon = 0$.

Many methods are available for identifying regular perturbation problems and for constructing their solutions. In this chapter, we derive some conditions that help to identify them as well as methods based on Taylor's theorem for constructing approximations to their solutions. Applications of these methods to oscillation problems are made in Chapter 6. Singular perturbation problems are studied using averaging and matching methods in Chapters 7 and 8, respectively.

The regular perturbation methods derived here are based on the Implicit Function Theorem. After introducing some other useful ideas, we derive methods for constructing solutions to initial value problems in both invertible and noninvertible cases.

5.1 Perturbation Expansions

Perturbation methods have played important roles in the development of mathematics and physics. Most of the problems studied here come from electrical circuits or mechanical systems, but the methods were developed for and used in a variety of applications.

In this chapter, we deal primarily with problems that can be solved using Taylor's formula, but methods using Pade approximations are also discussed. Recall that Taylor's formula shows how to approximate a smooth function by a polynomial, and it tells what error is made in the approximation. Pade's method uses rational function approximations, that is, those approximations made up of ratios of polynomials. Using rational functions often makes it possible to uncover information about singularities of a function and so to obtain approximations away from where power series converge.

A variety of notations are used to describe these procedures. Gauge functions are described first. Next, Taylor's and Pade's formulas are discussed. Since solutions of differential equations can be found using integral formulas, like the variation of constants formula, it is important to know Laplace's method and the method of stationary phase, which we also describe briefly in this section.

5.1.1 *Gauge Functions: The Story of o, O*

Suppose that f and g are smooth functions of ε for ε near zero, say $|\varepsilon| \leq \varepsilon_0$. We say that

$$f(\varepsilon) = O(g(\varepsilon)) \qquad \text{as } \varepsilon \to 0$$

if $f(\varepsilon)/g(\varepsilon)$ is bounded for all small ε. Thus, there is a constant K and a constant ε_1 such that

$$|f(\varepsilon)| \leq K|g(\varepsilon)|$$

for all $|\varepsilon| \leq \varepsilon_1$.

We say that

$$f(\varepsilon) = o(g(\varepsilon)) \qquad \text{as } \varepsilon \to 0$$

if

$$f(\varepsilon)/g(\varepsilon) \to 0 \qquad \text{as } \varepsilon \to 0.$$

Thus, given a tolerance η, there is a constraint δ such that $|f(\varepsilon)| \leq \eta|g(\varepsilon)|$ when $|\varepsilon| \leq \delta$. These are order-of-magnitude or *gauge relations* between functions, and they give some idea of the relative sizes between them for small values of the parameter ε. We say that

$$\sum_{n=0}^{N} c_n f_n(\varepsilon)$$

is an *asymptotic expansion* of $f(\varepsilon)$ at $\varepsilon = 0$ if the following are true:

1. The sequence $\{f_n\}$, $n = 0, \cdots, N + 1$, is a gauge sequence; that is, $f_n(\varepsilon) = o(f_{n-1}(\varepsilon))$ as $\varepsilon \to 0$ for $n = 1, \cdots, N + 1$, and

2.
$$f(\varepsilon) - \sum_{n=0}^{N} c_n f_n(\varepsilon) = O(f_{N+1}(\varepsilon)) \qquad \text{as } \varepsilon \to 0$$

Asymptotic expansions need not converge as $N \to \infty$, since convergence would imply that the function approximated is closely related to an analytic function. We have no reason to suspect *a priori* that solutions are that smooth. Asymptotic expansions do not uniquely determine functions, and on the other hand several different asymptotic sequences can be used to approximate a given function [see 5.1].

5.1.2   *Taylor's Formula*

A function $f(\varepsilon)$ that has $N + 1$ continuous derivatives at $\varepsilon = 0$, can be approximated by a polynomial of degree N:

$$f(\varepsilon) = f(0) + f'(0)\varepsilon + \cdots + f^{[N]}(0)\varepsilon^N/N! + O(\varepsilon^{N+1})$$

This is *Taylor's* expansion of f about $\varepsilon = 0$, and it gives a very useful approximation to f near $\varepsilon = 0$. The error is given explicitly by the formula

$$\frac{f^{[N+1]}(\eta)\varepsilon^{N+1}}{(N + 1)!}$$

where η is some point near zero, $|\eta| < \varepsilon$. Here $f_n = \varepsilon^n$, and so on.

5.1.3   *Pade's Approximations*

Pade approximations are rational functions that account for pole singularities, and they are frequently valid in domains larger than for power series expansions. It is often useful to construct an approximation to a function using Taylor's formula, and then derive Pade's approximation to Taylor's polynomial. This is obviously successful when we sum a geometric series. In general, an analytic function's Taylor series has radius of convergence R that is restricted to the largest circle not enclosing a singularity. If the singularity is a pole, then Pade's method can be used to continue the function outside this circle of convergence.

We say that the rational function

$$R_{m,n}(\varepsilon) = \frac{P_m(\varepsilon)}{Q_n(\varepsilon)}$$

where P_m is a polynomial of degree m and Q_n is one of degree n, is an $[m, n]$-*Pade approximation* to f if

$$f(\varepsilon) - R_{m,n}(\varepsilon) = O(\varepsilon^{m+n+1})$$

as $\varepsilon \to 0$. Constructing Pade's approximation, $R_{m,n}$ begins with a power series expansion of f about $\varepsilon = 0$; say

$$f(\varepsilon) = f_0 + f_1 \varepsilon + \cdots + f_M \varepsilon^M + O(\varepsilon^{M+1})$$

We specify m and n such that $m + n + 1 \leq M$, and we seek polynomials

$$P_m(\varepsilon) = a_0 + a_1 \varepsilon + \cdots + a_m \varepsilon^m$$

and

$$Q_n(\varepsilon) = b_0 + b_1 \varepsilon + \cdots + b_n \varepsilon^n$$

such that

$$f = \frac{P_m}{Q_n} + O(\varepsilon^{m+n+1})$$

Usually, we take $m = n - 1$, $2n = M$, and $b_0 = 1$. This last can be done without loss of generality if $Q(0) \neq 0$ since the numerator and denominator can be multiplied by an arbitrary constant. Since we use a ratio of polynomials of degrees m and n, we cannot expect to achieve greater accuracy than order $m + n + 1$, so we cross-multiply to get

$$Q_n f - P_m = O(\varepsilon^{m+n+1})$$

Equating coefficients of like powers on both sides gives

order $m + n$: $\quad b_0 f_{m+n} \quad + \cdots + b_n f_m \quad = 0$

order $m + n - 1$: $\quad b_0 f_{m-1+n} + \cdots + b_n f_{m-1} \quad = 0$

. . . .

order $m + 1$: $\quad b_0 f_{1+n} \quad + \cdots + b_n f_{m-n+1} = 0$

starting from order $m + n$. Here we set $f_j = 0$ if $j < 0$. This is a system of n equations for the n unknowns $b_1, \cdots, b_n$. Recall that $b_0 = 1$, so this is a nonhomogeneous problem.

We assume the following condition

Hypothesis H1. This system can be solved for $b_1, \cdots, b_n$.

Once the coefficients of Q_n are known, we can easily find the coefficients of P_m:

order 0: $\quad a_0 = f_0$

order 1: $\quad a_1 = b_1 f_0 + f_1$

. . . .

order m: $\quad a_m = \sum_{j=1}^{m} b_j f_{m-j} + f_m$

Thus, once the coefficients of the denominator are known, the coefficients of the numerator can be found explicitly.

There are many surprising and useful extensions of this method [see 5.2]. Consider the system of differential equations

$$\frac{dx}{dt} = A(\varepsilon)x + B(\varepsilon)$$

where A and B are matrices that are polynomials in ε, say of degrees m and n, respectively. A static state of this system must solve the equation

$$A(\varepsilon)x + B(\varepsilon) = 0$$

If $A(0)$ is invertible, then the unique static state is

$$x = -A^{-1}(\varepsilon)B(\varepsilon) \qquad \text{for } \varepsilon \text{ near } 0$$

which is a matrix of rational functions of ε. Therefore, Taylor's method is more restrictive than Pade's, for which a matrix form of Pade's method, $R_{m,n}(\varepsilon)$, gives the exact solution to this problem.

Two ways to proceed with perturbation problems are as follows:

Method I. Substitute Taylor's or Pade's expansion into the problem and equate coefficients of like powers of ε.

Method II. Derive equations for the coefficients in the approximation by successively differentiating the problem with respect to ε.

These two methods are used throughout the remainder of this book.

5.1.4   *Laplace's Methods*

The system of differential equations

$$\frac{dx}{dt} = A(\varepsilon)x + f(t)$$

has the solution

$$x(t) = e^{A(\varepsilon)t}x(0) + \int_0^t e^{A(\varepsilon)(t-s)}f(s)\,ds$$

Evaluating this formula can be difficult.

For example, if $A(\varepsilon) = 1/\varepsilon$, then the solution becomes large if $\varepsilon \to 0^+$ or if $t \to +\infty$. On the other hand, if $A(\varepsilon) = -1/\varepsilon$, then the first term on the right-hand side of this equation approaches a discontinuous limit as $\varepsilon \to 0$ namely, $x(0)$ if $t = 0$, but 0 if $t > 0$. The second term is more interesting, and the kernel in this integral is closely related to a Dirac delta function. Finally, if $A(\varepsilon) = i/\varepsilon$, then the integral becomes highly oscillatory, and the asymptotic evaluation of it is closely related to the Riemann-Lebesgue Lemma [5.1].

The first case ($A = 1/\varepsilon$) is divergent, and we do not deal with it further. The second case ($A = -1/\varepsilon$) is treated by using Laplace's method, and the third case ($A = i/\varepsilon$) can be evaluated using stationary phase methods. Since we are dealing primarily with approximations to solutions of differential equations, it is important to know these fundamental methods. In fact, they form the basis of the matching and averaging results, respectively, treated in later chapters. We begin with integration by parts, which is sometimes referred to as being the fundamental theorem of applied mathematics.

Integration by Parts

Consider the function of t and ε defined by the integral formula

$$\int_0^t \exp(is/\varepsilon) f(s)\, ds$$

where f is a continuously differentiable function. Integration by parts gives an easy way to determine how this function behaves as $\varepsilon \to 0$. In particular,

$$\int_0^t \exp\left(\frac{is}{\varepsilon}\right) f(s)\, ds = \frac{\varepsilon}{i} \exp\left(\frac{it}{\varepsilon}\right) f(t) - \frac{\varepsilon}{i} f(0) - \frac{\varepsilon}{i} \int_0^t \exp\left(\frac{is}{\varepsilon}\right) f'(s)\, ds$$

Integration by parts can be used to generate an asymptotic expansion of this function, but this short calculation shows that the integral is of order $O(\varepsilon)$ (as $\varepsilon \to 0$) for any fixed value of t.

Laplace's Integral Formulas

Consider the integral

$$h(\varepsilon) = \int_\alpha^\beta f(s) \exp\left(-\frac{g(s)}{\varepsilon}\right) ds$$

What is the dominant part of this integral as $\varepsilon \to 0^+$? Integration by parts is usually not useful in answering this question.

We suppose that $f(t)$ is continuous and $g(t)$ is a twice-differentiable real-valued function for $\alpha \le t \le \beta$. The place where g is smallest should give the fastest growing part of the kernel, so let us first suppose that g has a minimum at a point $\tau \in (\alpha, \beta)$. Then $g'(\tau) = 0$ and $g''(\tau) > 0$. As $\varepsilon \to 0$,

$$h(\varepsilon) \sim \int_{\tau-\delta}^{\tau+\delta} f(s) \exp\left\{-\frac{g(\tau) + g''(\tau)((s-\tau)^2/2)}{\varepsilon}\right\} ds$$

Setting $u^2 = g''(\tau)(s-\tau)^2/2\varepsilon$ gives

$$h(\varepsilon) \sim 2f(\tau) \exp\left[-\frac{g(\tau)}{\varepsilon}\right] \sqrt{\frac{2\varepsilon}{g''(\tau)}} \int_0^{\delta[g''(\tau)/2\varepsilon]^{1/2}} e^{-u^2}\, du$$

$$\sim 2f(\tau) \exp\left[-\frac{g(\tau)}{\varepsilon}\right] \sqrt{\frac{2\pi\varepsilon}{g''(\tau)}} \qquad \text{as } \varepsilon \to 0^+$$

When the minimum of g is attained at an end point of the interval, then similar calculations show that

$$h(\varepsilon) \sim \frac{f(\alpha)\exp[-g(\alpha)/\varepsilon]}{2}\sqrt{\frac{2\pi\varepsilon}{g''(\alpha)}} \qquad \text{if } g'(\alpha) = 0$$

$$\sim \frac{\varepsilon f(\alpha)\exp[-g(\alpha)/\varepsilon]}{g'(\alpha)} \qquad \text{if } g'(\alpha) > 0$$

if the left end point is a local minimum for g. Similarly for $\tau = \beta$.

The Method of Stationary Phase

Consider the integral formula

$$h(\varepsilon) = \int_\alpha^\beta f(s)\exp\left[\frac{i\phi(s)}{\varepsilon}\right]ds$$

where $|\varepsilon| \ll 1$ and the phase ϕ is a real-valued function of the real variable s. Stokes and Kelvin [5.1] claim that the major contributions to this integral come from the end points α and β and from the vicinity of the points where ϕ is stationary, that is, where $\phi' = 0$, and that to leading order in ε, the stationary points are more significant than the end points. This can be justified using the method of steepest descent, which we do not present here [see 5.1].

Suppose that ϕ is stationary at some point τ, say that $\phi'(\tau) = 0$ and $\phi''(\tau) > 0$. We suppose that the major part of the integral comes from near τ, so we write

$$h(\varepsilon) \sim \int_{\tau-\varepsilon}^{\tau+\varepsilon} f(t)\exp\left[\frac{i\phi(t)}{\varepsilon}\right]dt$$

As before

$$h(\varepsilon) \sim \sqrt{\frac{2\pi\varepsilon}{\phi''(\tau)}}f(\tau)\exp\left[\frac{i\phi(\tau)}{\varepsilon} + \frac{i\pi}{4}\right] \qquad \text{as } \varepsilon \to 0^+$$

We use these calculations later in Chapter 7.

5.2 Regular Perturbations of Initial-Value Problems

What conditions ensure that a problem is regular; for example, when will Taylor's formula be successful in finding approximate solutions? Such a result follows.

5.2.1 *Regular Perturbation Theorem*

Given smooth data f and ξ in E^N, we consider the initial value problem

$$\frac{dx}{dt} = f(t, x, \varepsilon) \qquad x(0) = \xi(\varepsilon)$$

which depends on a real parameter ε. Suppose the following conditions:

Hypothesis H2. For $\varepsilon = 0$, this system has a unique solution on some finite interval $0 \leq t \leq T$. We call this solution $x_0(t)$, and it satisfies the equations

$$\frac{dx_0}{dt} = f(t, x_0, 0) \qquad x_0(0) = \xi(0)$$

Hypothesis H3. f and $\xi(\varepsilon)$ are smooth functions of their variables for $0 \leq t \leq T$, x near x_0, and ε near zero. Specifically, we suppose that f and ξ are $N + 1$ times continuously differentiable in all variables, so

$$\xi(\varepsilon) = \xi_0 + \xi_1 \varepsilon + \cdots, \text{ etc.}$$

We have the following theorem:

Regular Perturbation Theorem. *Let conditions H2 and H3 be satisfied. Then for sufficiently small ε the perturbed problem has a unique solution, and it is $N + 1$ times differentiable with respect to ε. Moreover, this solution can be expanded in a Taylor expansion*

$$x(t, \varepsilon) = x_0(t) + x_1(t)\varepsilon + \cdots + x_N(t)\varepsilon^N + O(\varepsilon^{N+1})$$

where the error estimate holds as $\varepsilon \to 0$ uniformly for $0 \leq t \leq T$.

5.2.2 *Proof of the Regular Perturbation Theorem*

The proof of this result involves three steps. The usual existence theory for ordinary differential equations ensures that a unique solution of the problem exists for $0 \leq t \leq T$ [5.3]. Denote this by $x(t, \varepsilon)$. Since this solution is $N + 1$ times differentiable with respect to ε, it can be expanded in a Taylor polynomial up to order $O(\varepsilon^{N+1})$. The first step involves finding this expansion. The difference between the real solution and Taylor's approximation is called the remainder. The second step involves deriving an equation for the remainder, and third step involves finding an *a priori* estimate of the remainder.

Step 1. We use Method II for finding the expansion of the solution:

$$x(t, \varepsilon) = x_0(t) + x_1(t)\varepsilon + \cdots$$

where

$$x_0(t) = x(t, 0)$$

$$x_1(t) = \left(\frac{\partial x(t, \varepsilon)}{\partial \varepsilon} \right) \qquad \text{at } \varepsilon = 0, \quad \text{etc.}$$

Therefore,

$$\frac{dx_0}{dt} = f(t, x_0, 0) \qquad\qquad x_0(0) = \xi_0$$

$$\frac{dx_1}{dt} = f_x(t, x_0, 0)x_1 + f_\varepsilon(t, x_0, 0) \quad x_1(0) = \xi_1$$

$$\cdots\cdots\cdots\cdots\cdots\cdots\cdots\cdots\cdots\cdots\cdots$$

$$\frac{dx_m}{dt} = f_x(t, x_0, 0)x_m + \{\cdots\}_m \qquad x_m(0) = \xi_m$$

Here f_x denotes the Jacobian matrix having components $(\partial f_i/\partial x_j)$, f_ε denotes the partial derivatives of the components of f with respect to ε, and the terms $\{\cdots\}_m$ denote a combination of terms depending on $x_0, \cdots, x_{m-1}$.

The first equation's solution is given by condition H2. The remaining equations for $m = 1, \cdots, M$, are forced linear equations, and their solutions can be obtained from the variation of constants formulas. Thus, the expansion of $x(t, \varepsilon)$ is uniquely determined.

Step 2. The remainder is defined to be

$$R(t, \varepsilon) = x(t, \varepsilon) - E(t, \varepsilon, N)$$

where E denotes the expansion just found for $x(t)$:

$$E(t, \varepsilon, N) = \sum_{j=0}^{N} x_j(t)\varepsilon^j$$

This function satisfies the equation

$$\frac{dR}{dt} = f(t, R + E, \varepsilon) - \frac{dE}{dt}$$

An equivalent way to think of E is that it was found by substituting the form

$$E(t, \varepsilon, N) = \sum_{j=0}^{N} x_j(t)\varepsilon^j$$

into the equation and equating like powers of ε. Thus,

$$\frac{dE}{dt} = f(t, E, \varepsilon) + O(\varepsilon^{N+1})$$

and the equation for R becomes

$$\frac{dR}{dt} = (f_x(t, x_0, 0) + O(\varepsilon))R + o(R) + O(\varepsilon^{N+1}), \qquad R(0) = O(\varepsilon^{N+1})$$

The second step involves showing that this equation has a unique solution. This follows from the method of successive approximations done in Chapter 4.

Step 3. Finally, we must estimate $|R|$. It follows from Gronwall's inequality that $R = O(\varepsilon^{N+1})$. This completes the proof of the theorem. ∎

The proof of this result is typical of proofs of expansion approximations. First, the approximation is constructed on faith. Then the difference between it and a real solution of the problem is defined. An equation for this difference is then analyzed, usually using contraction mapping arguments. This establishes the existence of a solution to the remainder equation. Finally, the solution of the remainder equation is estimated, usually using Gronwall's inequality.

5.2.3   *Example of the Regular Perturbation Theorem*

Consider the equation

$$\frac{dx}{dt} = -\frac{x}{1 + \varepsilon} \qquad x(0) = \cos \varepsilon$$

The regular perturbation theorem guarantees that the solution of this problem is a smooth function of ε at $\varepsilon = 0$. Therefore, it has a Taylor expansion about $\varepsilon = 0$, say

$$x(t, \varepsilon) = x_0(t) + x_1(t)\varepsilon + \cdots + x_N(t)\varepsilon^N + O(\varepsilon^{N+1})$$

These coefficients can be determined directly from the equation by successively differentiating it with the result being a sequence of differential equations for the coefficients: (Method II)

$$\frac{dx_0}{dt} = -x_0 \qquad\qquad x_0(0) = 1$$

$$\frac{dx_1}{dt} = -x_1 + x_0 \qquad\qquad x_1(0) = 0$$

$$\frac{dx_2}{dt} = -x_2 + x_1 - x_0 \qquad x_2(0) = -1, \quad \text{etc.}$$

Therefore,

$$x_0(t) = \exp(-t)$$
$$x_1(t) = t \exp(-t)$$
$$x_2(t) = \exp(-t)(t^2/2 + t - 1), \quad \text{etc.}$$

This calculation illustrates the direct Taylor method; of course, Pade's method can work better for this problem.

5.2.4   *Regular Perturbations for $0 \le t < \infty$*

Two interesting examples are

$$\frac{dx}{dt} = -x^3 + \varepsilon x$$

and

$$\frac{dx}{dt} = \varepsilon x$$

Both of these are regular perturbation problems for t restricted to a finite interval, but the regular perturbation theorem breaks down for each on infinite time intervals. In the first case, $x(t) \to \sqrt{\varepsilon}$ as $t \to \infty$, which is not a smooth function of ε at $\varepsilon = 0$, even though the equations are smooth. In the second, $x(t, \varepsilon)$ is an entire function of t and ε, but the solution is $\exp(\varepsilon t)$, which does not converge uniformly for $0 \le t < \infty$ as $\varepsilon \to 0$.

What conditions ensure that the approximation found in the regular perturbation theorem is valid uniformly for all $0 \le t < \infty$? Consider the initial value problem

$$\frac{dx}{dt} = f(t, x, \varepsilon) \qquad x(0) = \xi(\varepsilon)$$

Suppose first that the problem has a solution when $\varepsilon = 0$.

Hypothesis H4. The zero-order problem

$$\frac{dx_0}{dt} = f(t, x_0, 0) \qquad x_0(0) = \xi(0)$$

has a unique solution for $0 \le t < \infty$.

Next, suppose that the data are smooth near this solution.

Hypothesis H5. ξ and f are smooth functions of their arguments for ε near zero, $0 \le t < \infty$ and for (t, x) near $(t, x_0(t))$.

Finally, suppose that the linearized problem is exponentially asymptotically stable (EAS).

Hypothesis H6. The fundamental solution of the linear problem

$$\frac{dy}{dt} = f_x(t, x_0(t), 0)y, \qquad y(s) = I$$

satisfies

$$|y(t)| \le K \exp(-\alpha(t - s))$$

for all $0 \le t < \infty$ where K and α are some positive constants.

With these assumptions, we have the following theorem.

Regular Perturbation Theorem on [0, ∞). *Let conditions H4, H5, and H6 be satisfied. Then for sufficiently small ε, the full problem has a unique solution and it exists for $0 \le t < \infty$. Moreover, if*

$$E(t, \varepsilon, N) = \sum_{j=0}^{N} x_j(t)\varepsilon^j$$

is the expansion derived in the Regular Perturbation Theorem, then

$$x(t, \varepsilon) - E(t, \varepsilon, N) = O(\varepsilon^{N+1})$$

where the error estimate holds uniformly for $0 \le t < \infty$.

The proof of this result proceeds exactly the same as in the finite interval case, except for the last step. In step 3, we must use the estimate provided by condition H6 in Gronwall's inequality to realize a uniform bound on the function $(x - E)\varepsilon^{-M-1}$ that is valid for $0 \le t < \infty$. This is the same argument as used in the Linear Stability Theorem in Chapter 3, and it is not reproduced here.

Example. Uniformly valid regular perturbation expansion. Consider the equation

$$\frac{dx}{dt} = Ax + \varepsilon f(x) \qquad x(0) = \xi$$

where $f(0) = 0$ and A is an asymptotically stable matrix (i.e., its eigenvalues have strictly negative real parts.)
    Then

$$x(t, \varepsilon) - \exp(At)x(0) = O(\varepsilon)$$

as $\varepsilon \to 0$ uniformly for $0 \le t < \infty$.

5.3 Modified Perturbation Methods

Consider now an autonomous (i.e., time invariant) system of equations

$$\frac{dx}{dt} = f(x, \varepsilon)$$

Static states of the problem are determined by solving the equation

$$f(x, \varepsilon) = 0.$$

We studied this in Chapter 4 where the implicit function theorem was derived.
    Recall that we assumed that

Hypothesis H1. There is a solution for $\varepsilon = 0$, say x_0^*:

$$f(x_0^*, 0) = 0.$$

Hypothesis H2. f is a smooth function of x and ε for x near x_0^* and ε near zero.

Hypothesis H3. The Jacobian

$$\det(\partial f/\partial x)(x_0^*, 0) \neq 0.$$

Or,

Hypothesis H3'. The Jacobian matrix has rank $r < N$.

We refer to problems satisfying conditions H3 as being nondegenerate and those satisfying H3' as being degenerate.

5.3.1 *Nondegenerate Static-State Problems Revisited*

Under conditions H1, H2, and H3, there is a unique solution of the perturbed equation near x_0^*, say $x^*(\varepsilon)$, for ε near zero. Moreover, this solution is a smooth function of ε. It follows that $x^*(\varepsilon)$ can be expanded in Taylor's expansion about $\varepsilon = 0$:

$$x^*(\varepsilon) = x_0 + \varepsilon x_1 + \cdots + \varepsilon^N x_N + O(\varepsilon^{N+1})$$

Since the coefficients in this expansion are derivatives of x^* with respect to ε, we can derive equations for them directly from the defining equation:

$$f(x_0, 0) = 0$$

$$f_x(x_0, 0)x_1 + f_\varepsilon(x_0, 0) = 0$$

and so on. Since x_0 is given and the Jacobian condition is satisfied, each of these equations has a unique solution.

5.3.2 *Modified Perturbation Theorem*

In Chapter 4, the implicit function theorem was used to study problems where the linear part is not invertible, but we had to derive and deal with a complicated set of bifurcation equations. In this section, we derive a useful perturbation technique for constructing the bifurcation equations directly.

Consider the static state problem

$$f(x, \varepsilon) = 0$$

where $f(0, 0) = 0$ and f is a smooth function near $(0, 0)$ in E^{N+1}. Thus, we suppose that $x_0 = 0$. Further, suppose that

Hypothesis H3'. The Jacobian matrix

$$A = (\partial f/\partial x)(0, 0)$$

has rank r and nullity $q = N - r$. Moreover, the null space of this matrix is spanned by vectors $\{\phi_1, \cdots, \phi_q\}$ and the null space of its adjoint is spanned by vectors $\{\psi_1, \cdots, \psi_q\}$ where

$$\psi_j \cdot \phi_k = \delta_{j,k}$$

$$\phi_j \cdot \phi_k = \delta_{j,k}$$

for j, $k = 1, \cdots, q$. Here $\delta_{j,k}$ is Kronecker's delta function; it is equal to 1 if $j = k$ and is zero otherwise.

Finally, we rewrite the equation as

$$f(x, \varepsilon) = Ax + G(x, \varepsilon) = 0$$

where $G(x, \varepsilon) = f(x, \varepsilon) - Ax = O(\varepsilon + |x|^2)$ for x and ε near zero.

We are now in a position to define the modified problem.

Modified Perturbation Problem

Given small values of $c_1, \cdots, c_q$ and ε, find functions $\lambda_j(c_1, \cdots, c_q, \varepsilon)$ for $j = 1, \cdots, q$, and $x(c_1, \cdots, c_q, \varepsilon)$ such that

$$Ax + G(x, \varepsilon) + \sum_{j=1}^{q} \lambda_j \psi_j = 0$$

$$\phi_k \cdot x = c_k \qquad \text{for } k = 1, \cdots, q$$

The following result shows that this system is solvable, and that the bifurcation equations derived in Chapter 4 are found by setting $\lambda_1 = 0, \cdots, \lambda_q = 0$.

Modified Perturbation Theorem. *Under conditions H1, H2, and H3′, the modified perturbation problem has a unique solution. That is, there are unique functions $\lambda_j(c_1, \cdots, c_q, \varepsilon)$ and $w(c_1, \cdots, c_q, \varepsilon)$ that are defined for $c_1, \cdots, c_q, \varepsilon$ near zero such that*

1. *The functions $x = \sum_{j=1}^{q} c_j \phi_j + w$ and $\lambda_1, \cdots, \lambda_q$ satisfy the modified perturbation problem.*
2. *These functions have at least $N + 1$ continuous derivatives with respect to $c_1, \cdots, c_q$ and ε. In particular, these functions can be expanded using Taylor's formula*

$$w = w_{c_1} c_1 + \cdots + w_{c_q} c_q + w_\varepsilon \varepsilon + \text{h.o.t.}$$

$$\lambda_j = \lambda_{jc_1} c_1 + \cdots + \lambda_{jc_q} c_q + \lambda_{j\varepsilon} \varepsilon + \text{h.o.t.}$$

This result is quite useful in studying resonance problems. Of course, it is a clever restatement of the degenerate Implicit Function Theorem [5.4]

Proof of the Modified Perturbation Theorem. The proof of this result is accomplished by showing that the Implicit Function Theorem can be used to solve the modified problem. Writing

$$x = \sum_{j=1}^{q} c_j \phi_j + w$$

we have

$$Aw + \Psi\lambda = -G(x, \varepsilon)$$

The columns of A span the set $\ker(A^{tr})^{\perp}$ and the columns of the matrix Ψ span the $\ker(A^{tr})$. Therefore, the linear part on the left-hand side of this equation is a matrix having N linearly independent columns, and so it is invertible. Therefore, the modified problem can be solved using the non-degenerate form of the Implicit Function Theorem. In particular, there is a unique solution for w and λ, which are smooth functions of $c_1, \cdots, c_q$ and ε. In fact, introduce a change of basis

$$x = \sum_{j=1}^{q} c_j \phi_j + \sum_{j=1}^{N-q} w_j \phi_{j+q}$$

where the vectors $\phi_1, \cdots, \phi_q$, span the $\ker(A)$ and the vectors $\phi_{q+1}, \cdots, \phi_{q+r}$ complete this set to a basis of E^N. Then the modified problem becomes

$$\lambda_j = -\psi_j \cdot G\left(\sum_{j=1}^{q} c_j \phi_j + w, \varepsilon \right)$$

$$Aw = G\left(\sum_{j=1}^{q} c_j \phi_j + w, \varepsilon \right) - \sum_{j=1}^{q} \lambda_j \psi_j$$

Since the linear problem for λ and w has full rank, the Implicit Function Theorem (nondegenerate form) ensures that these equations can be solved for small solutions λ and w as functions of $c_1, \cdots, c_q$ and ε for $|c|$ and $|\varepsilon|$ near zero. ∎

Thus, there is a unique solution for x and λ as functions of c and ε. Once this is done, the original problem can be solved by requiring that

$$\lambda_j(c, \varepsilon) = 0 \qquad \text{for } j = 1, \cdots, q$$

These are equivalent to the bifurcation equations derived in Chapter 4.

5.3.3   *Example: $q = 1$*

Approximations of the modified perturbation solution (i.e., x and the functions λ_j) can be found using Taylor's formula: We take $q = 1$ to illustrate this method. Then

$$x = x_\varepsilon \varepsilon + x_c c + \cdots$$

$$\lambda = \lambda_\varepsilon \varepsilon + \lambda_c c + \cdots$$

The first result is

$$Ax_\varepsilon + G_\varepsilon(0, 0) + \lambda_\varepsilon \psi = 0$$

$$x_\varepsilon \cdot \phi = 0$$

Thus, $\lambda_\varepsilon = -\psi \cdot G_\varepsilon(0,0)$ and

$$Ax_\varepsilon + G_\varepsilon(0,0) - (\psi \cdot G_\varepsilon(0,0))\psi = 0$$

This equation has a unique solution that is orthogonal to ϕ. Next,

$$Ax_c + \lambda_c\psi = 0$$
$$x_c \cdot \phi = 1$$

Therefore, $\lambda_c = 0$ and $x_c = \phi$.

In this way, we can continue to construct Taylor's expansion of each of these functions. We return to the original problem by forcing the modified problem to be the same as the original one:

$$\lambda(c, \varepsilon) = 0$$

This equation can be studied using Newton's polygons. The modified method applies in the exchange of stabilities problem in Chapter 4, and we use it in constructing nonlinear oscillations in Chapter 6.

5.4 Exercises

5.1. Show that x and $x + \exp(-1/x)$ have the same asymptotic expansions as $x \to 0^+$.

5.2. Show that

$$\frac{1}{1+x} \sim -\sum (-x)^n,$$
$$\sim -\sum (x-1)x^{-2n}$$
$$\sim -\sum (x^2 - x + 1)(-x)^{-3n}$$

as $x \to \infty$ where the sums are taken for $n = 0, 1, 2, 3, 4, \cdots$.

5.3. a. Solve the equation

$$(1 - \varepsilon)f(\varepsilon) = 1$$

for f using two different methods: Pade approximants and Taylor's expansions. Compare the answers you obtain in each case.

b. Construct the $[1,1]$-Pade approximation to the function

$$f(\varepsilon) = \left(\frac{1 + \varepsilon/2}{1 + 2\varepsilon}\right)^{-1/2}$$

Construct Taylor's approximation of f up to order 2. Compare your answers for $\varepsilon = 1$ and for $\varepsilon = \infty$.

5.4. Calculate an asymptotic approximation to the integral

$$\int_a^1 \exp\frac{is^2}{\varepsilon} \cos s \, ds$$

as $\varepsilon \to 0$ for the two cases $a = 0$ and $a = -1$.

5.5. Show that $R = O(\varepsilon^{M+1})$ in the proof of the regular perturbation theorem in Section 5.2.2.

5.6. Use the Ansatz $x = \exp[P(\varepsilon)t/Q(\varepsilon)]$ to solve the problem

$$\frac{dx}{dt} = -\frac{x}{1+\varepsilon}, \qquad x(0) = \cos\varepsilon$$

in the example in Section 5.2.3 for $|\varepsilon| \ll 1$.

5.7. Apply the modified perturbation method to the exchange of stabilities problem in Sections 4.3.1 and 5.3.3.

6
Forced Oscillations

Forcing a nonlinear oscillation can have complicated consequences, especially if the forcing period is near the oscillation's period. The simplest question is under what conditions will there be a periodic solution? The answers for externally forced systems are quite different from those for feedback systems that are found in Sections 3.5, 7.4, 7.5, and 8.4.

Problems considered in this chapter have the general form

$$\frac{dx}{dt} = f(t, x, \varepsilon)$$

where $x, f \in E^N$ and f is a smooth function of its variables. We suppose that the unperturbed problem

$$\frac{dx}{dt} = f(t, x, 0)$$

has a periodic solution $x = p(t)$, say with period T. Conditions on the data will be found under which the perturbed problem ($\varepsilon \neq 0$) will have a periodic solution that is close to $p(t)$ and has period near T or some rational number multiple of T.

The system

$$\frac{dx}{dt} = f(t, x, \varepsilon)$$

where f has least period $T > 0$, say $f(t + T, \cdots) = f(t, \cdots)$ for all t, is referred to as an *externally forced system*. When f does not depend explicitly on t, we refer to it as being a *feedback system*.

6.1 Resonance

When we try to construct expansions of forced oscillations, we soon discover difficulties if the forcing frequency is near a natural frequency of the system. Two results, one describing nonresonance and the other applicable to resonant cases are presented in the first section of this chapter. The modified perturbation method is applied to these cases, and the results are compared to other perturbation and iteration schemes in the case of Duffing's equation. In the course of this, we discover that chaotic behavior can occur when Duffing's equation is externally forced, and we investigate some aspects of this at the end of the chapter.

The equation

$$\frac{d^2x}{dt^2} + \mu^2 x = \varepsilon \cos t$$

is useful for introducing some basic ideas. It can be solved explicitly. If $\mu^2 \neq 1$, then

$$x(t) = A \cos(\mu t + \phi) + \frac{\varepsilon \cos t}{\mu^2 - 1}$$

where A and f are arbitrary constants. The first term gives the general solution of the free problem. If $\mu^2 = 1$, then

$$x(t) = A \cos(t + \phi) + \frac{\varepsilon t \sin t}{2}$$

the last term is referred to as being secular since it is not bounded.

Since we are interested in finding solutions that have the same period as the forcing function (viz., 2π), we specify that x satisfy the periodicity conditions

$$x(t + 2\pi, \varepsilon) = x(t, \varepsilon)$$

for all $t \geq 0$.

There are three cases:

1. The problem is *nonresonant* or regular when μ is not an integer. In this case, the equation has a unique solution of period 2π; namely,

$$x(t) = \frac{\varepsilon \cos t}{\mu^2 - 1}$$

2. *Subresonance* occurs when $\mu^2 = n^2$ for some integer $n^2 \neq 1$. In this case, the solution formula shows that both terms in the solution have period 2π, but the first, which corresponds to the general solution of the homogeneous equation, has least period $2\pi/n$. The second term is a particular solution of the equation.
3. *Resonance* occurs when $\mu^2 = 1$. In this case, there is no solution of the equation having period 2π.

The situation here is quite similar to Fredholm's Alternative for solving systems of linear equations. Recall that when solving the linear system

$$Lx = f$$

where L is a matrix and $\mathbf{x}$ and $\mathbf{f}$ are vectors, we found that there is a solution for $\mathbf{x}$ if and only if the forcing function $\mathbf{f}$ is orthogonal to the null space of L^{tr}, the adjoint of L. We follow the same approach for solving equations where now L represents a differential operator and $\mathbf{f}$ is a function.

The concept of orthogonality for vectors extends naturally to functions. Let f and g be two complex valued functions defined on an interval $0 \le t \le T$. Let $t_j = jT/N$ for $j = 0, 1, 2, \cdots, N$, and $f_j = f(t_j)$ and $g_j = g(t_j)$. The vectors $\mathbf{f} = (f_0, \cdots, f_N)$ and $\mathbf{g} = (g_0, \cdots, g_N)$ have inner product

$$\mathbf{f} \cdot \mathbf{g} = \frac{1}{N} \sum_{j=0}^{N} f_j g_j^*$$

where g_j^* denotes the complex conjugate of g_j. If N is large, this is approximately the integral

$$\mathbf{f} \cdot \mathbf{g} = \frac{1}{T} \sum_{j=0}^{N} f_j g_j^* \frac{T}{N} \sim \frac{1}{T} \int_0^T f(s) g^*(s)\, ds \equiv (f, g)$$

where we define the inner-product of two square integrable functions $f(t)$ and $g(t)$ to be

$$(f, g) = \frac{1}{T} \int_0^T f(t) g^*(t)\, dt$$

We say that f and g are orthogonal over $[0, T]$ if $(f, g) = 0$.

Note that if f is a square-integrable complex-valued function, then (f, f) exists and so we can define

$$|f| = \sqrt{(f, f)}$$

We have seen that these integral formulas are natural extensions to functions of geometric ideas for vectors.

For example, taking the inner product of both sides of our differential equation with the general solution of the unforced problem ($\varepsilon = 0$), namely with $\cos(\mu t + \phi)$, we have

$$\left(\cos(\mu t + \phi), \frac{d^2 x}{dt^2} + \mu^2 x \right) = (\cos(\mu t + \phi), \varepsilon \cos t)$$

Integration by parts shows that the left-hand side of this equation is zero. Therefore, a necessary condition that this equation have a solution is that

$$0 = (\cos(\mu t + \phi), \varepsilon \cos t)$$

for any constant ϕ. In this example, this condition is also sufficient to ensure solvability.

How can these cases be determined directly from the equation? The unperturbed, or free, equation is

$$\frac{d^2 x}{dt^2} + \mu^2 x = 0$$

Since we will be forcing this system with period 2π, solutions of this equation that have period 2π are the important ones. In the resonance case, the forcing is among these. In the subresonance case, the forcing function is orthogonal to these, and in the regular case, there are no such solutions to the free problem.

In our analogy with Fredholm's Alternative, the matrix L is replaced by a differential operator

$$Lx = \frac{d^2 x}{dt^2} + \mu^2 x$$

defined for all square-integrable functions $x(t)$ for which the second derivative makes sense and which satisfy the periodicity conditions

$$x(t) = x(t + 2\pi)$$

for all t. This operator is self-adjoint in the sense that $(Lx, y) = (x, Ly)$.

6.1.1   *Formal Perturbation Expansion of Forced Oscillations*

Let us try to construct the solution of the general problem

$$\frac{dx}{dt} = f(t, x, \varepsilon) \qquad x(t + 2\pi) = x(t)$$

by a regular perturbation expansion beginning with a known periodic solution when $\varepsilon = 0$, say $p(t)$, without worrying about validity. Let our Ansatz, or initial guess for the solution, be

$$x(t) = p(t) + \varepsilon x_1 + \varepsilon^2 x_2 + \cdots$$

This gives

$$\frac{dp}{dt} = f(t, p, 0)$$

$$\frac{dx_1}{dt} = f_x(t, p, 0)x_1 + f_\varepsilon(t, 0, 0)$$

and so on, where the subscripts denote differentiations with respect to ε and the components of x.

The solution for x_1 is given by the formula

$$x_1(t) = Q(t)\exp(Rt)x_1(0) + \int_0^t Q(t)\exp[R(t - s)]Q^{-1}(s)f_\varepsilon(s)\,ds$$

where $Q(t)\exp(Rt)$ is the fundamental solution of the linear part given by Floquet's theory. x_1 is periodic if $x_1(T) = x_1(0)$. Thus, we have a formula for the initial data of periodic solutions:

$$[I - \exp(RT)]x_1(0) = \int_0^T \exp[R(T - s)]Q^{-1}(s)f_\varepsilon(s)\,ds$$

As before, there are three cases:

1. The matrix $I - e^{RT}$ is invertible.
2. The matrix is not invertible, but the right-hand side is orthogonal to its adjoint null vectors (subresonance).
3. The matrix $I - e^{RT}$ is not invertible and the right-hand side is not orthogonal to its adjoint null vectors (resonance).

In case 3, there is no solution for $x_1(0)$ and so the formal perturbation Ansatz breaks down. In case 2 there is a unique particular solution, plus an arbitrary linear combination of null vectors of the matrix. This requires the special accounting supplied by the modified perturbation method. In the first case, there is a unique solution for $x_1(0)$, hence a unique periodic solution of the system.

This calculation illustrates the general situation: Specific results are described in the sections on nonresonance and resonance, which follow.

6.1.2 Nonresonant Forcing

It is possible to give conditions that ensure nonresonance and to completely describe perturbation solutions in that case. We refer to this result as being the Nonresonance Theorem.

Consider the forced system

$$\frac{dx}{dt} = f(t, x, \varepsilon)$$

where f is a vector of smooth functions that are periodic in t, say with least period T. We suppose the following:

Hypothesis H1. There is a solution $x = p(t)$ of the unperturbed equation

$$\frac{dx}{dt} = f(t, x, 0)$$

that has period T.

Hypothesis H2. $x, f \in E^N$. f is a smooth function of its variables for (t, x) near $(t, p(t))$ and ε sufficiently small. Moreover, f has least period $T > 0$ in t:

$$f(t + T, x, \varepsilon) = f(t, x, \varepsilon)$$

for all $t \geq 0$.

The problem here is to determine solutions of this system that are periodic having the same period as the forcing

$$x(t + T, \varepsilon) = x(t, \varepsilon)$$

for all $t \geq 0$.

If $\psi(t, a, \varepsilon)$ is the solution of the differential equation that satisfies the initial condition

$$\psi(0, a, \varepsilon) = p(0) + a$$

then fixed points of the mapping

$$p(0) + a \to \psi(T, a, \varepsilon)$$

define periodic solutions of the original problem. Therefore, we consider the equation

$$\psi(T, a, \varepsilon) - p(0) - a = 0$$

We can apply the Implicit Function Theorem to this equation to construct solutions for $a = a(\varepsilon)$, and so eventuallly construct periodic solutions of the forced problem.

The Jacobian matrix of this equation for a is

$$\frac{\partial \psi}{\partial a}(T, 0, 0) - I$$

where I is the N-dimensional identity matrix. If this matrix is nonsingular, then we can find a unique solution for $a = a(\varepsilon)$. If the matrix is singular, then we can use the modified perturbation method to study solutions.

The Jacobian matrix is usually not easy to determine, because the following calculation must be done. The function $u = \partial \psi / \partial a$ is an $N \times N$ matrix that solves the linear problem

$$\frac{du}{dt} = f_x(t, p(t), 0)u(t) \qquad u(0) = I$$

Since this is a linear equation having periodic coefficients, Floquet's theory shows that $u(t)$ has the form

$$u(t) = P(t)\exp(Rt)$$

where $P(t)$ has period T in t and R is a constant matrix. Therefore, $u(T) = e^{RT}$. Recall that the matrix R is usually difficult to find.

It follows that if all eigenvalues of R lie off the imaginary axis in the complex plane, then the regular perturbation method works directly. The result is a uniquely determined function $a(\varepsilon)$. This proves the following theorem

Nonresonance Theorem. *Let conditions* H1 *and* H2 *above be satisfied. Further, suppose that $x = 0$ is the only solution of the linear problem*

$$\frac{dx}{dt} = f_x(t, p(t), 0)x \qquad x(t + 2\pi) = x(t)$$

then for sufficiently small ε, the full problem has a unique periodic solution $x = x^*(t, \varepsilon)$. If f is $M + 1$ times differentiable with respect to its variables, then *x* can be constructed using a regular perturbation expansion:*

$$x^*(t, \varepsilon) = p(t) + x_1(t)\varepsilon + \cdots + x_M(t)\varepsilon^M + O(\varepsilon^{M+1})$$

where the coefficients are uniquely determined by a hierarchy of linear problems. x* is determined by the initial conditions

$$x^*(0, \varepsilon) = p(0) + a(\varepsilon)$$

where a is determined in the preceding argument.

Example. Duffing's equation with nonresonant forcing. Here we try to construct a 2π-periodic solution of the equation

$$\frac{d^2x}{dt^2} + \mu^2 x + \varepsilon f(x) = g(t)$$

where f is a smooth function, μ is a fixed constant, and ε is a small parameter. The forcing function g is 2π-periodic and has a Fourier series

$$g(t) = \sum_{m=-\infty}^{m=\infty} g_m e^{imt}$$

This equation includes Duffing's equation for which

$$f(x) = \alpha x + \beta x^3.$$

We suppose here that μ is not an integer. The unperturbed problem is obtained by setting $\varepsilon = 0$:

$$\frac{d^2x}{dt^2} + \mu^2 x = g(t)$$

and it has a unique periodic solution that is given by the Fourier series

$$p(t) = \sum_{m=-\infty}^{\infty} g_m \frac{e^{imt}}{\mu^2 - m^2}$$

In this case, according to the Regular Perturbation Theorem, we can find x as a smooth function of ε. Therefore, we try to find the solution of the equation in the form of a Taylor's expansion

$$x(t, \varepsilon) = p(t) + \varepsilon x_1(t) + \cdots$$

We obtain a system of equations for the other coefficients

$$\frac{d^2x_1}{dt^2} + \mu^2 x_1 = f(x_0)$$

$$\frac{d^2x_2}{dt^2} + \mu^2 x_2 = -\frac{\partial f}{\partial x}(x_0)x_1$$

and so on. These equations are to be solved subject to the periodicity conditions

$$x_j(t + 2\pi) = x_j(t) \qquad \text{for } j = 1, 2, \cdots$$

for all t. Obviously, each of these problems has a unique solution, and the Fourier series of each can be found.

6.1.3 Resonant Forcing

Problems created by resonant forcing are quite similar to those encountered in the noninvertible case of the Implicit Function Theorem. We can reduce the problem to a set of bifurcation equations, but the solutions of these may be quite difficult to find.

We consider here systems in the form

$$\frac{dx}{dt} = Ax + \varepsilon G(t, x, \varepsilon)$$

where $x, G \in E^N$. We say the forcing function (G) is *resonant* if it has period T and if the matrix $I - \exp(AT)$ is not invertible. In this case, the hypotheses of the Nonresonance Theorem are not satisfied since the linear problem

$$dx/dt = Ax$$

has a (nonzero) solution of period T.

The following conditions are used in this section:

Hypothesis H1. G is a smooth function for $-\infty < t < \infty$, with x near zero and $|\varepsilon| \ll 1$, and $G(t + T, x, \varepsilon) = G(t, x, \varepsilon)$ for all t and some least $T > 0$. Moreover, $G(t, x, 0) = O(|x|)^2$ for x near zero uniformly for $0 \leq t \leq T$.

Hypothesis H2. The matrix A has the form

$$A = \begin{pmatrix} R & 0 \\ 0 & D \end{pmatrix}$$

where R is a $K \times K$ diagonal matrix, $R = \text{diag}(r_1, \cdots, r_K)$, with $\exp(RT) = I$, and the matrix D is such that $\exp(DT) - I$ is invertible.

This system is not as general as could be considered here, but it enables us to deal with the fundamental solution $\exp(At)$ explicitly. More general equations require evaluation of the Floquet exponent matrix, which we have discussed earlier.

The existence of periodic solutions can be studied like before: We seek initial values a such that the solution $x = \phi(t, a, \varepsilon)$ has period T. As before,

$$\phi(0, a, \varepsilon) = a$$

and we want to solve

$$\phi(T, a, \varepsilon) = a.$$

for a.

ϕ solves the equivalent integral equation

$$\phi(t, a, \varepsilon) = e^{At}a + \varepsilon \int_0^t e^{A(t-s)}G(s, \phi, \varepsilon)\, ds$$

The periodicity condition becomes

$$(I - e^{AT})a = \varepsilon \int_0^T e^{A(T-s)}G(s, \phi, \varepsilon)\, ds$$

Setting $\hat{a} = \mathrm{col}(a_1, \cdots, a_K)$ and $\tilde{a} = \mathrm{col}(a_{K+1}, \cdots, a_N)$, and similarly for G, say $\hat{G} = (G_1, \cdots, G_K)$ and $\tilde{G} = \mathrm{col}(G_{K+1}, \cdots, G_N)$, then we have

$$0 = \int_0^T e^{-Rs}\hat{G}(s, \phi(s, a, \varepsilon), \varepsilon)\, ds$$

$$(I - e^{DT})\tilde{a} = \varepsilon \int_0^T e^{D(T-s)}\tilde{G}(s, \phi(s, a, \varepsilon), \varepsilon)\, ds$$

The Implicit Function Theorem guarantees that for each small ε and for each choice of $\hat{a}$ near zero, there is a unique solution for $\tilde{a}$, say

$$\tilde{a} = \tilde{a}(\hat{a}, \varepsilon)$$

Substituting this solution into the first K equations gives the bifurcation equations, which must be solved for $\hat{a}$. These results are summarized in the following theorem.

Resonant Forcing Theorem. *Let conditions H1 and H2 hold. Then for each small $\hat{a}$ in E^K and ε there exists a unique solution $\tilde{a} = \tilde{a}(\hat{a}, \varepsilon)$ in E^{N-K} of*

$$(I - e^{DT})\tilde{a} = \varepsilon \int_0^T e^{D(T-s)}\tilde{G}(s, \phi(s, a, \varepsilon), \varepsilon)\, ds$$

Moreover, for each small solution $\hat{a}$ of the bifurcation equations

$$0 = \int_0^T e^{-Rs}\hat{G}(s, \phi(s, \tilde{a}(\hat{a}, \varepsilon), \varepsilon), \varepsilon)\, ds$$

there is a periodic solution

$$x = \phi(t, a, \varepsilon)$$

of the original problem where $a = \mathrm{col}(\hat{a}, \tilde{a}(\hat{a}, \varepsilon))$.

This result is less satisfactory than the Nonresonance Theorem because it takes us only to the bifurcation equations, and they are in a form that requires knowledge of $\phi(t, a, \varepsilon)$. The modified perturbation method enables us to construct the bifurcation equations, although there may be no solutions to them, or they might not be solvable using available methods.

6.1.4 *Modified Perturbation Method for Forced Oscillations*

Suppose that the conditions of the resonant forcing theorem are satisfied. Then we can pursue periodic solutions as we did static states, using the modified perturbation method.

First, we augment the problem by introducing K new variables $\lambda_1, \cdots, \lambda_K$, and K new constraints $c_1, \cdots, c_K$:

$$\frac{dx}{dt} = Ax + \varepsilon G(t, x, \varepsilon) + \varepsilon \sum_{j=1}^{K} \lambda_j e^{r_j t} \mathbf{e}_j$$

where $\mathbf{e}_j$ is the jth standard basis unit

$$\mathbf{e}_j = (\delta_{1,j}, \cdots, \delta_{N,j})$$

and r_j is the jth eigenvalue of R. Note that because of our choice of basis in the original problem, the last term can be written as

$$\varepsilon \exp(At)\lambda$$

where λ is the vector

$$\lambda = \text{col}(\lambda_1, \cdots, \lambda_K, 0, \cdots, 0).$$

Thus, the augmented problem is

$$\frac{dx}{dt} = Ax + \varepsilon G(t, x, \varepsilon) + \varepsilon \exp(At)\lambda$$

The K additional constraints are specified by the projections

$$(x, \exp(r_j t)\mathbf{e}_j) = c_j$$

for $j = 1, \cdots, K$. Finally, we specify that x has period T by requiring that

$$x(t + T) = x(t)$$

for all t.

Second, we construct Taylor expansions for x and λ as functions of ε and $c = (c_1, \cdots, c_K)$, for these quantities near zero. Each of these problems has a unique solution, and the results are functions

$$x = x(t, c, \varepsilon) \qquad \text{and} \qquad \lambda = \lambda(c, \varepsilon)$$

The third step gets us back to the original problem. We require that

$$\lambda(c, \varepsilon) = 0$$

These are the bifurcation equations, and when they are satisfied, $x(t, c, \varepsilon)$ is a solution of the original problem that has period T.

Rather than present this method as a general algorithm, we consider an important example in Section 6.2.8.

6.1.5 *Justification of the Modified Perturbation Method*

The program prescribed by the Modified Perturbation Method has three steps:

1. Add K variables $\lambda_1, \cdots, \lambda_K$ and K constraints $a_1, \cdots, a_K$ to the problem by

$$\frac{dx}{dt} = Ax + \varepsilon G(t, x, \varepsilon) + \varepsilon \exp(At)\lambda$$

$$(x, \exp(rt)e_j) = c_j$$

where $\lambda = \mathrm{col}(\lambda_1, \cdots, \lambda_K, 0, \cdots, 0)$ and e_j is the jth standard basis unit (i.e., a vector with 1 in the jth position and 0 elsewhere). Note that the right-hand side of the differential equation is T-periodic in t, and we are to solve this system for x and λ subject to the constraints and the periodicity condition

$$x(t + T) = x(t)$$

for all t.

2. Construct x and λ as power series in ε and the components of $c = \mathrm{col}(c_1, \cdots, c_K, 0, \cdots, 0)$. The coefficients in these expansions are determined uniquely.

3. The equations

$$\lambda_j(c, \varepsilon) = 0 \qquad j = 1, \cdots, K,$$

must be solved for c as functions of ε, say $c = c(\varepsilon)$. For each such solution the function $x(t, c(\varepsilon), \varepsilon)$ is a T-periodic solution of the original problem.

We now show that this approach is equivalent to the one in the previous section. In fact, let us denote the solution of the modified differential equations that satisfies $x(0) = b$ by

$$y(t, b, \varepsilon)$$

This function satisfies the integral equation

$$y(t, b, \varepsilon) = e^{At}b + \varepsilon \int_0^t e^{A(t-s)} G(s, y, \varepsilon)\, ds + \varepsilon t e^{At}\lambda$$

y has period T (i.e., $y(0) = y(T)$) if and only if b satisfies the equation

$$(I - e^{AT})b = \varepsilon \int_0^T e^{A(T-s)} G(s, y, \varepsilon)\, ds + \varepsilon T e^{AT}\lambda$$

Note the presence of a secular term in the equation for $y(t, b, \varepsilon)$; these are discussed from another point of view in Section 6.2.3. We write $b = \mathrm{col}(\hat{b}, \tilde{b})$ where the first subvector has K components, etc. For each choice of $\hat{b}$ near zero and $|\varepsilon| \ll 1$, there is a unique solution for $\tilde{b} = \tilde{b}(\hat{b}, \varepsilon)$ of the equations

$$(I - e^{DT})\tilde{b} = \varepsilon \int_0^T e^{D(T-s)} \tilde{G}(s, y, \varepsilon)\, ds$$

We now replace b by $\mathrm{col}(\hat{b}, \tilde{b}(\hat{b}, \varepsilon))$. With this, the projection constraints take the form

$$c_j = (y, \exp(r_j t)e_j) = \hat{b}_j + \varepsilon H_j(\hat{b}, \varepsilon)$$

where

$$H_j = \frac{1}{T} \int_0^T \int_0^t e^{-r_j s} \hat{G}_j\, ds\, dt + \frac{T}{2}\lambda_j$$

for $j = 1, \cdots, K$. The Implicit Function Theorem guarantees that

$$c = \hat{b} + \varepsilon H(\hat{b}, \varepsilon)$$

defines a one-to-one relation between c and $\hat{b}$. In particular, we can solve for $\hat{b}$ say with the result

$$\hat{b} = c + \varepsilon h(c, \varepsilon)$$

The functions λ_j are defined by the formulas

$$\lambda_j(c, \varepsilon) = -\frac{1}{T} \int_0^T e^{-r_j s} \hat{G}_j\, ds$$

which results from direct integration of the differential equation for y. Finally, values of c and ε, for which $\lambda = 0$, result in a solution y that has period T, but since it now satisfies the same equation as ϕ, we also have that for such c and ε,

$$y(t, b(c, \varepsilon), \varepsilon) = \phi(t, a, \varepsilon)$$

Thus, the results of the Resonant Forcing Theorem and the Modified Perturbation Method agree.

6.2   Duffing's Equation

The examples in this section illustrate several interesting aspects of resonant forcing. These are all based on Duffing's equation, which is among the simplest nonlinear conservation equations. Several methods different from the Modified Perturbation Method have been devised to study resonantly forced systems. Two particularly important ones are Duffing's iterative procedure and the Poincare-Linstedt perturbation procedure. Rather than describing these methods in general, we consider an important typical case: We study the response of Duffing's equation to small resonant forcing where the modified perturbation method, Duffing's iteration, and the Poincare-Linstedt methods can all be used. One or another of these methods can work for many other cases.

We consider in this section the problem

$$\frac{d^2x}{dt^2} + x = \varepsilon(-\alpha x - \beta x^3 + C\cos\omega t)$$

Resonance occurs when the forcing frequency ω is near 1.

6.2.1 *Modified Perturbation Method*

Consider Duffing's equation when the forcing frequency is near resonance, say

$$\omega^2 = 1 + \mu\varepsilon + O(\varepsilon^2)$$

With the change of variable $\omega t \to t$, the equation becomes

$$(1 + \varepsilon\mu)\frac{d^2x}{dt^2} + x = \varepsilon(-\alpha x - \beta x^3 + C\cos t)$$

According to the modified perturbation method, we augment the equation by introducing λ and η as shown next

$$(1 + \varepsilon\mu)\frac{d^2x}{dt^2} + x = \varepsilon(-\alpha x - \beta x^3 + C\cos t + \lambda\cos t + \eta\sin t)$$

and we specify the constraints

$$(x, \cos t) = a \quad \text{and} \quad (x, \sin t) = b$$

The Modified Perturbation Method is quite clumsy in this case. As a shortcut, we ignore the fact that a and b are small, and we construct x, λ, and η in powers of ε alone:

$$\frac{d^2x_0}{dt^2} + x_0 = 0$$

$$(x_0, \cos t) = a \quad \text{and} \quad (x_0, \sin t) = b$$

$$\frac{d^2x_1}{dt^2} + x_1 = -\alpha x_0 - \beta x_0^3 + C\cos t - \mu\frac{d^2x_0}{dt^2} + \lambda_0\cos t + \eta_0\sin t$$

$$(x_1, \cos t) = 0 \quad \text{and} \quad (x_1, \sin t) = 0$$

and so on. Solving the first problem gives

$$x_0(t) = a\cos t + b\sin t$$

It will turn out that $b = 0$, so we use this information now to simplify the calculations. In other that the second problem be solvable for a function having period 2π, we must have the coefficient of $\cos t$ equal to zero on the right hand side of the equation:

$$-\alpha a - 3\beta\frac{a^3}{4} + C + a\mu + \lambda_0 = 0$$

this formula relates the forcing frequency (ω) to the amplitude of response (a). For further details see Refs. 6.1 and 6.2.

The bifurcation equation is found by setting $\lambda = 0$. To leading order in ε, this is

$$3\beta\frac{a^3}{4} + (\alpha - \mu)a - C = 0$$

Note that there is a cusp singularity at $a = 0$, $\alpha - \mu = 0$, $C = 0$, as shown in Figure 6.1.

6.2.2 Duffing's Iterative Method

Duffing introduced an iterative method [6.3] for approximating the periodic solutions of the equation

$$\frac{d^2x}{dt^2} + x = \varepsilon(-\alpha x - \beta x^3 + C\cos\omega t)$$

Begin with an initial guess suggested by the forcing, namely,

$$x_0(t) = A\cos\omega t$$

and consider the iteration

$$\frac{d^2x_{n+1}}{dt^2} + x_{n+1} = \varepsilon(-\alpha x_n - \beta x_n^3 + C\cos\omega t)$$

for $n = 0, 1, \cdots$. The equation for x_1 is

$$\frac{d^2x_1}{dt^2} + x_1 = \varepsilon(-\alpha A\cos\omega t - \beta(A\cos\omega t)^3 + C\cos\omega t)$$

The right-hand side can be simplified by writing

$$(\cos\omega t)^3 = \frac{3\cos\omega t}{4} + \frac{\cos 3\omega t}{4}$$

Then

$$\frac{d^2x_1}{dt^2} + x_1 = \varepsilon\left(\left(-\alpha A - \frac{3\beta A^3}{4} + C\right)\cos\omega t - \frac{A^3\beta}{4}\cos 3\omega t\right)$$

Integrating this (assuming $\omega \neq 1$) gives

$$x_1 = \frac{\varepsilon(-\alpha A - 3\beta A^3/4 + C)}{1 - \omega^2}\cos\omega t - \frac{A^3\beta}{4(1 - 9\omega^2)}\cos 3\omega t$$

Duffing's idea is that if the initial guess is a good one, then the coefficient of $\cos\omega t$ in this solution must be close to A. Thus,

$$(1 - \omega^2)A = \varepsilon\left(-\alpha A - \frac{3\beta A^3}{4} + C\right)$$

If $\omega^2 = 1 + \mu\varepsilon$, then

$$\frac{3\beta A^3}{4} + (\alpha - \mu)A - C = 0$$

This is identical to the result obtained in the preceding section using the Modified Perturbation Method.

Once A is determined in this way, it follows that the amplitude of $\cos\omega t$ in each of the subsequent iterations will also be A.

6.2.3 Poincare-Linstedt Method

In the two preceding sections, the forcing frequency ω was given and the response's amplitude was determined. It is possible to reverse the procedure by specifying the amplitude of the response and asking what frequencies give a 2π-periodic oscillation having this amplitude.

Again, we consider the equation

$$\frac{d^2x}{dt^2} + x = \varepsilon(-\alpha x - \beta x^3 + C\cos\omega t)$$

We make the change of variables $\omega t \to t$ with the result being

$$\omega^2 \frac{d^2x}{dt^2} + x = \varepsilon(-\alpha x - \beta x^3 + C\cos t)$$

Determine ω and x such that the periodicity condition

$$x(t + 2\pi) = x(t)$$

is satisfied as well as the initial conditions

$$x(0) = A \qquad x'(0) = 0$$

which specifies that x has A as a (local) maximum value.

We begin by expanding x and ω in power series

$$x = x_0(t) + x_1(t)\varepsilon + \cdots$$

$$\omega = \omega_0 + \omega_1\varepsilon + \cdots$$

Setting $\varepsilon = 0$ in the problem gives

$$\omega_0^2 \frac{d^2x_0}{dt^2} + x_0 = 0$$

$$x_0(t + 2\pi) = x_0(t)$$

$$x_0(0) = A, \qquad x_0'(0) = 0$$

The periodicity condition requires that

$$\omega_0 = 1 \qquad \text{and} \qquad x_0(t) = A\cos t$$

Differentiating the equation with respect to ε and setting $\varepsilon = 0$ (Method II) gives

$$\omega_0^2 \frac{d^2 x_1}{dt^2} + x_1 = -\alpha A \cos t - \beta A^3 \cos^3 t + C \cos t - 2\omega_0 \omega_1 \frac{d^2 x_0}{dt^2}$$

or

$$\frac{d^2 x_1}{dt^2} + x_1 = \left(C - \alpha A - \frac{3\beta A^3}{4} + 2\omega_1 A \right) \cos t - \frac{\beta A^3}{4} \cos 3t$$

We choose ω_1 to cancel the remaining part of the coefficient of $\cos t$ in the right-hand side:

$$\omega_1 = \frac{\alpha A + 3\beta A^3/4 - C}{2A}$$

Therefore, we have

$$\omega = 1 + \varepsilon \frac{\alpha A + 3\beta A^3/4 - C}{2A} + O(\varepsilon^2)$$

If we are interested in the frequency near 1, we write

$$\omega = 1 + \varepsilon \eta/2$$

which is the expansion of $\omega^2 = 1 + \varepsilon \eta$ for ε near zero, and we get

$$\frac{3\beta A^3}{4} + (\alpha - \eta)A - C + O(\varepsilon) = 0$$

This relation between forcing frequency and response amplitude agrees with the ones derived in the preceding sections.

The Poincare-Linstedt method is quite similar to the Modified Perturbation Method, but its point of view is reversed as noted earlier. The method is sometimes referred to as being *the* perturbation method, and sometimes it is called the *Method of Suppressing Secular Terms*. Secular terms refer to non-periodic solutions that grow slowly. They can be removed by choosing constants so that coefficients of terms causing resonance are zero, as we chose ω_1 here. This is very much in the spirit of the Modified Perturbation Method, and its validity is established in our discussion of Averaging Methods in Chapter 7.

6.2.4 Frequency-Response Surface

Each of the methods used above leads to a frequency-response relation of the form

$$A^3 - \frac{4(\mu - \alpha)A}{3\beta} - \frac{4C}{3\beta} = 0$$

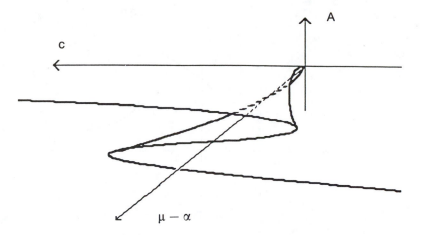

FIGURE 6.1. Cusp frequency-response surface.

although Duffing's iteration method involves the least work in deriving this formula. This formula defines a surface in $AC\mu$ space, and it is depicted in Figure 6.1 where it is assumed that all coefficients in this equation are positive.

Fixing the forcing amplitude C determines a cross-section of this surface. Several of these are depicted in Figure 6.2 where we plot the response ampitude $|A|$.

The restoring force $\alpha x + \beta x^3$ is due to the potential energy

$$U(x) = \frac{\alpha x^2}{2} + \frac{\beta x^4}{4}$$

In analogy with a mechanical spring, the case when $\beta > 0$ is referred to as being a *hard spring*, since increasing extension (x) increases the restoring force. $\beta < 0$ is a *soft spring*, since the restoring force eventually decreases with increasing extension. These diagrams show that there is a nonunique response of the system for most parameter values. More insight on this comes from considering Duffing's equation with damping, as in Section 6.2.6.

6.2.5 *Subharmonic Responses of Duffing's Equation*

The periodic solutions found in the preceding section are harmonics of the forcing; that is, they have the same period as the forcing function. Many other periodic solutions of Duffing's equation are excited by periodic forcing. Solutions having least period being some integer multiple of the forcing period are common; these are called *subharmonics* (lower frequency). Solutions where the least period is the forcing period divided by an integer are called *higher harmonics*, because their frequencies are integer multiples of the base frequency.

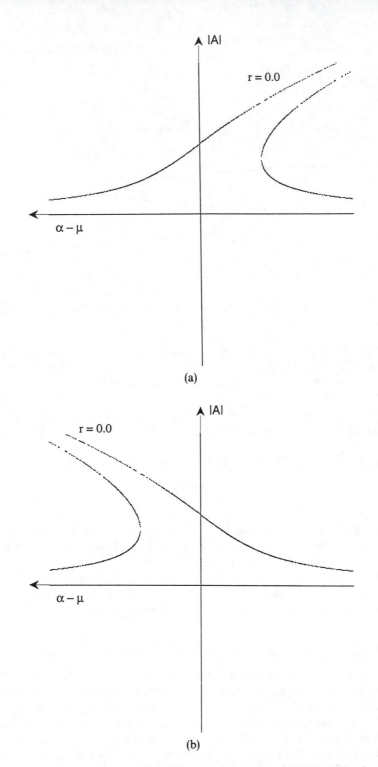

FIGURE 6.2. Frequency-response for (a) hard springs, where $\beta > 0$ and (b) soft springs where $\beta < 0$. $C > 0$ is fixed.

Subharmonics can be found by using Duffing's iterative method and a suitable initial guess that picks them out. For example, let

$$x_0 = a\cos\frac{t}{3} + A\cos t$$

be our initial guess for the 1/3 harmonic of the equation

$$\omega^2\frac{d^2x}{dt^2} = -\alpha x - \beta x^3 + F\cos t$$

where ω is the forcing frequency. Our guess has a least period of 6π. Iterating beginning with x_0, and equating coefficients of the iterate to those of x_0 give

$$a\left(\alpha - \frac{\omega^2}{9}\right) + \frac{3\beta}{4}(a^3 + a^2A + 2aA^2) = 0$$

$$A(\alpha - \omega^2) + \frac{\beta}{4}(a^3 + 6a^2A + 3A^3) = F$$

Let us consider this system in the linear case where $\beta = 0$. There are nonzero solutions for a and A only if $\omega = \sqrt{\alpha}$ or $\omega = 3\sqrt{\alpha}$. There are nearby solutions of these equations if β is near zero and ω is near these two values.

The important point here is that the linear problem gives the hint of where to look for subharmonics in the space of all possible parameters (α, β, and ω) appearing in the problems. If the forcing frequency is a rational multiple of the free frequency of the linear problem, then we might expect a comparable oscillation for the nonlinear problem. This may or may not happen, but it usually gives a starting point.

6.2.6 Damped Duffing's Equation

Now consider the equation

$$\frac{d^2x}{dt^2} + r\frac{dx}{dt} + \alpha x + \beta x^3 = a\cos\omega t + b\sin\omega t$$

where $r > 0$ is the damping coefficient. Let us use Duffing's method to study this equation. We begin with an initial guess

$$x_0 = A\cos\omega t$$

and define the iteration

$$\frac{d^2x_1}{dt^2} = -r\frac{dx_0}{dt} - \alpha x_0 - \beta x_0^3 + a\cos\omega t + b\sin\omega t$$

Integrating this equation gives

$$x_1(t) = \frac{(3\beta A^3/4) + \alpha A - a}{\omega^2}\cos\omega t - \frac{b + \omega Ar}{\omega^2}\sin\omega t + \text{h.f.t.}$$

where h.f.t. denotes higher frequency terms. Setting this equal to the initial guess gives

$$b = -A\omega r$$

$$a = \frac{3\beta A^3}{4} + \alpha A - A$$

Let us rewrite the forcing term as an amplitude C and a phase lag ϕ, where $C^2 = a^2 + b^2$ and $\phi = -\tan^{-1}(b/a)$. With this, $a\cos\omega t + b\sin\omega t = C\cos(\omega t + \phi)$, and

$$C^2 = (\omega A r)^2 + \left(\frac{3\beta A^3}{4} + (\alpha - \omega^2)A\right)^2$$

Now, the cusp singularity is at $C^2 = (\omega A r)^2$. If r is small, the response shown in Figure 6.3 is similar to that in Figure 6.2, with the important exception that there is a unique small amplitude oscillation when $|\omega^2 - a| \gg 1$.

Figure 6.3 explains some interesting physical behavior of electrical circuits [see 6.4]. If we begin with a soft spring oscillator with $\alpha > \omega^2$, then the system oscillates with amplitude shown on the right side of the vertical axis in Figure 6.3. As the forcing frequency ω increases, the oscillation remains on the top branch until the fold is reached. It then jumps to the lower branch; that is, to a small amplitude oscillation. This kind of discontinuous behavior is often observed in what appear to be very smooth systems. We say that there is a *hysteretic* response of the system. For example, after this experiment establishes the oscillation on the lower branch far to the left of the vertical axis, we can begin to decrease ω. As we do this, the oscillation remains small in amplitude until the lower fold is reached near the vertical axis when the oscillation jumps to higher amplitude.

6.2.7   *Computer Simulation of Duffing's Equation*

The behavior described in Figure 6.3 can be illustrated by direct computation of the solutions to Duffing's equation. Figure 6.4 depicts a portion of the phase plane for (free) Duffing's equation. The center is surrounded by a continuum of periodic orbits that extend to the separatrices that join the two saddle points. Plotted are values $\{(x(2\pi n), y(2\pi n))\}$ for many choices of n and for many choices of initial conditions $x(0)$, $y(0)$.

Shown in Figure 6.5 is Poincare's mapping for a damped equation that is externally forced

$$0.81\frac{d^2x}{dt^2} + 0.05\frac{dx}{dt} + x - x^3 = \cos t$$

This plot is obtained by selecting an initial point, say $x(0) = \xi$, $dx/dt(0) = \eta$, solving the equation, and plotting the solution point each time t hits an integer multiple of 2π. The plot is generated by repeating this simulation for many initial points.

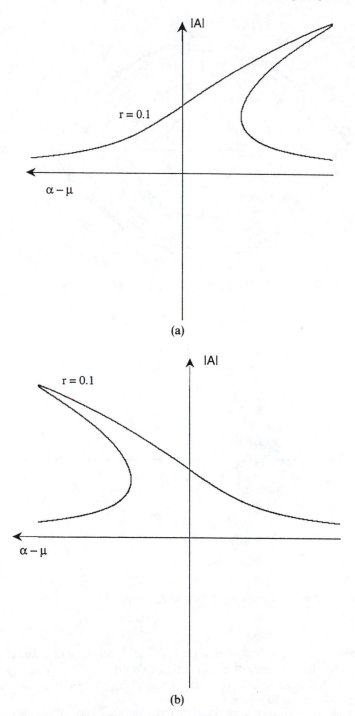

FIGURE 6.3. Frequency-response for damped Duffing equation, in the hard spring (a) and soft spring (b) cases. In both cases $0 < r < C/\omega A$. $\mu = \omega^2$ here.

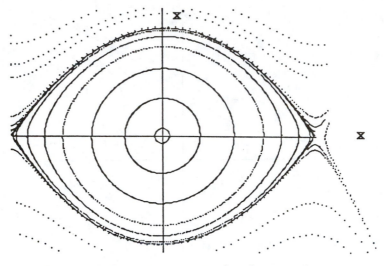

FIGURE 6.4. Phase plane for $0.81(d^2x/dt^2) + x - x^3 = 0$.

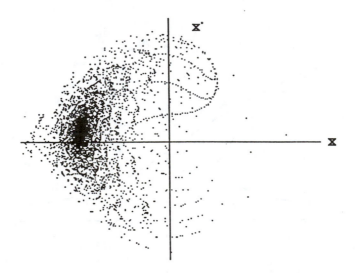

FIGURE 6.5. Poincare's mapping for a forced, damped Duffing equation ($r = 0.05$).

We found here a unique stable periodic solution of period 2π. In Figure 6.6, we solve the same equation but with weaker damping. In this case, we see two stable periodic solutions, one corresponding to each of the top and bottom branches in Figure 6.3.

The basins of attraction of these two oscillations are separated by a very complicated curve that is in some sense random. To understand this, we turn to a discussion in Section 6.3 of fractal curves and chaotic behavior.

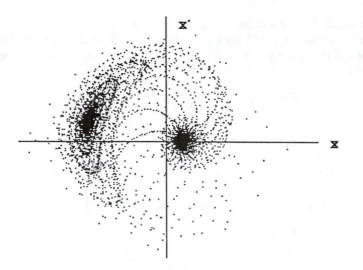

FIGURE 6.6. Poincare's mapping for the equation

$$0.81\frac{d^2x}{dt^2} + 0.02\frac{dx}{dt} + x - x^3 = \cos t$$

6.2.8   Duffing's Equation with Subresonant Forcing

The resonant forcing theorem applies as it did in Section 6.1.3 when we replace εG by

$$g(t) + \varepsilon F(t, x, \varepsilon)$$

where g is subresonant. This is illustrated here with a calculation using the modified perturbation method. Consider the equation

$$\frac{d^2x}{dx^2} + \mu^2 x = C\cos t - \varepsilon(\alpha x + \beta x^3)$$

where $\mu \neq \pm 1$ is an integer. The forcing function $\cos t$ has period 2π and the two linearly independent solutions of the homogeneous equation

$$\frac{d^2x}{dt^2} + \mu^2 x = 0$$

also have period 2π, but the forcing is orthogonal to them.

The modified perturbation method gives a way to construct periodic solutions of this equation. We introduce new variables λ and η and constraints a and b as follows:

$$\frac{d^2x}{dt^2} + \mu^2 x = C\cos t - \varepsilon(\alpha x + \beta x^3) + \varepsilon\lambda\cos\mu t + \varepsilon\eta\sin\mu t$$

$$(x, \cos\mu t) = a \qquad \text{and} \qquad (x, \sin\mu t) = b$$

$$x(t + 2\pi) = x(t) \qquad \text{for all } t$$

We successively differentiate this problem with respect to ε, a, and b, and so derive a sequence of problems for expansion coefficients (the subscripts here denote differentiations and numbered subscripts denote differentiations with respect to ε)

$$\frac{d^2 x_0}{dt^2} + \mu^2 x_0 = C \cos t$$

$$(x_0, \cos \mu t) = 0 \qquad \text{and} \qquad (x_0, \sin \mu t) = 0$$

$$\frac{d^2 x_1}{dt^2} + \mu^2 x_1 = -\alpha x_0 - \beta x_0^3 + \lambda_0 \cos \mu t + \eta_0 \sin \mu t$$

$$(x_1, \cos \mu t) = 0 \qquad \text{and} \qquad (x_1, \sin \mu t) = 0$$

$$\frac{d^2 x_a}{dt^2} + \mu^2 x_a = 0$$

$$(x_a, \cos \mu t) = 1 \qquad \text{and} \qquad (x_a, \sin \mu t) = 0$$

$$\frac{d^2 x_b}{dt^2} + \mu^2 x_b = 0$$

$$(x_b, \cos \mu t) = 0 \qquad \text{and} \qquad (x_b, \sin \mu t) = 1$$

$$\frac{d^2 x_{a\varepsilon}}{dt^2} + \mu^2 x_{a\varepsilon} = -(\alpha + 3\beta x_0^2) x_a + \lambda_a \cos \mu t + \eta_a \sin \mu t$$

$$(x_{a\varepsilon}, \cos \mu t) = 0 \qquad \text{and} \qquad (x_{a\varepsilon}, \sin \mu t) = 0$$

$$\frac{d^2 x_{b\varepsilon}}{dt^2} + \mu^2 x_{b\varepsilon} = -(\alpha + 3\beta x_0^2) x_b + \lambda_b \cos \mu t + \eta_b \sin \mu t$$

$$(x_{b\varepsilon}, \cos \mu t) = 0 \qquad \text{and} \qquad (x_{b\varepsilon}, \sin \mu t) = 0, \quad \text{etc.}$$

The solution of the first problem is

$$x_0(t) = \frac{C \cos t}{(\mu^2 - 1)}$$

Note that λ_0 is zero if $\mu \neq 3$, but otherwise it is not zero. The other solutions are

$$\eta_0 = 0$$

$$x_a(t) = \cos \mu t / \pi$$

$$x_b(t) = \sin \mu t / \pi$$

$$\lambda_a = \frac{\alpha}{\pi} + \frac{3\beta C^2}{2(\mu^2 - 1)^2}$$

$$\eta_b = \frac{\alpha}{\pi} + \frac{3\beta C^2}{2(\mu^2 - 1)^2}$$

$$\lambda_b = 0$$

$$\eta_a = 0, \quad \text{etc.}$$

We have enough here to demonstrate two important points. First, if $\mu = 3$ and $C \neq O(\varepsilon)$, then $\lambda_0 \neq 0$, and we have no hope of solving the bifurcation equation

$$\lambda = 0$$

for small a, b, and ε. Therefore, the method does not produce a solution because of interference of higher harmonics. Second, if $\mu = 3$ and $C = O(\varepsilon)$ or if $\mu \neq 3$, then the Jacobian matrix of the bifurcation equations is

$$\frac{\partial(\lambda, \eta)}{\partial(a, b)} = \frac{\alpha}{\pi} + \frac{3\beta C^2}{2(\mu^2 - 1)^2}$$

If this quantity is not zero, there are unique functions

$$a = a(\varepsilon) \qquad \text{and} \qquad b = b(\varepsilon)$$

that solve the bifurcation equations

$$\lambda(a, b, \varepsilon) = 0$$

$$\eta(a, b, \varepsilon) = 0$$

and the resulting function $x(t, a(\varepsilon), b(\varepsilon), \varepsilon)$ is a 2π-periodic solution of Duffing's equation.

6.3 Boundaries of Basins of Attraction

When two stable modes of oscillation coexist, like in the cases shown in Figure 6.3 and 6.6 where the response curve has doubled back on itself, there comes an important question. Given an initial point, what will happen to its solution? There are three possibilities: the solution can approach one of the two stable oscillations or it can remain on the boundary separating their basins of attraction. This can be important in meeting design specifications and in determining stability of a designed system. Unfortunately, the boundary between two basins of attraction can be quite complicated. In particular, there can be exquisitely sensitive dependence of solutions on their initial positions. This fact was known to Poincare and others in the last century and a number of researchers since then have studied aspects of this phenomenon. It is closely related to the chaotic behavior described in Chapter 2.

Newton's method gives an excellent starting point for a discussion of basins of attraction. Next, some computer simulations are used to illustrate chaotic behavior in Duffing's and van der Pol's oscillators. The boundaries between basins of attraction can be very complicated, and our discussion of fractals provides some insight into how bizarre things can become. Finally, a computer simulation to estimate the nature of the boundary in specific cases is described.

6.3.1 *Newton's Method and Chaos*

Newton's method gives us one of the best available algorithms for finding the zeros of a function. It is a recursive scheme whose iterates move in the space in which solutions lie. It also presents an interesting introduction to chaos.

Recall that if $f(x)$ is a smooth real-valued function, and if we seek the values of x for which $f(x) = 0$, then we can proceed by successive linear extrapolations. Let us make an initial guess, say x_0, and define a sequence of approximations $\{x_n\}$ by the algorithm

$$x_{n+1} = x_n - \frac{f(x_n)}{f'(x_n)}$$

as long as this makes sense. This formula defines Newton's method. If $f'(x^*) \neq 0$, where x^* is the zero near x_0, then this sequence will converge to x^*, and it will usually do so at a quadratic rate. Thus, quite rapid convergence results when the method works.

Sarkovski studied what happens when the condition $f'(x^*) \neq 0$ is not satisfied [6.5]. He derived the following interesting results, which were described in Section 2.5.1. A computer simulation was designed and carried out [6.6] to find the zero of the function

$$f(x) = \left(\frac{|x - r + 1|}{x}\right)^{1/(r-1)}$$

where $r > 1$. This function has a discontinuous derivative at its zero $x = r - 1$ (i.e., $f(r - 1) = 0$). If we derive Newton's algorithm for this, we get for $1 < r \leq 4$

$$x_{n+1} = rx_n(1 - x_n)$$

For these values the sequence of iterates lies in the unit interval $0 \leq x \leq 1$, which we studied in section 2.5.1.

6.3.2 *Computer Examples*

The one-dimensional mappings do not elucidate the basin-of-attraction boundary problem, but three other examples are presented here to do this. Consider a mapping of the plane E^2 into itself,

$$F: \mathbf{x} = (x, y) \to (f(x, y), g(x, y)) = F(\mathbf{x})$$

Suppose that F has two fixed points, say $F(\mathbf{a}) = \mathbf{a}$ and $F(\mathbf{b}) = \mathbf{b}$. Furthermore, suppose that each of these fixed points is stable. We define the basins of attraction of these points to be

$$B(\mathbf{a}) = (\mathbf{x} \in E^2 : F^n(\mathbf{x}) \to \mathbf{a} \text{ as } n \to \infty)$$

where $F^n(x) = F(F^{n-1}(x))$. We define $B(\mathbf{b})$ similarly.

The boundary between these two sets might be simple or very complicated. A question of interest is given a point in E^2, does it lie in $B(\mathbf{a})$ or in $B(\mathbf{b})$? Points that lie away from the boundary are decidable, but those that lie on it are not. The following examples illustrate several interesting facts.

6.3.2.1   The Roots of $z^3 - 1$

First, let us consider iterates of Newton's method for the complex valued function $f(z) = z^3 - 1$. There are three zeros of this function, one at each of the three cubic roots of unity: $z = 1, z = -1/2 \pm i\sqrt{3}/2$. Iterates of this mapping were studied by Fatou, Julia, and Kinney and Pitcher among many others. It is known that the three domains of attraction, one each for the three roots, are hopelessly intertwined. In fact, at any point that bounds any two of these domains, there are points of the third domain arbitrarily close by. The set bounding these three basins of attraction is now referred to as a *Julia set* (see Figure 6.7.) [6.9, 6.10, 6.11, 6.12].

6.3.2.2   Duffing's Equation

Figure 6.5 shows a case where there is a unique periodic response. In Figure 6.6, two stable responses coexist. The two basins of attraction in this case are separated by a complicated curve. We return later to finding out how complicated this curve can be.

6.3.2.3   Van der Pol's Equation

Van der Pol's equation has the form

$$\frac{d^2x}{dt^2} + k(x^2 - 1)\frac{dx}{dt} + x = B\cos t$$

If k is large and $B = kA$, then all of the responses lie in an annulus created around the knee-shaped curve. [6.7, 6.8].

The solutions can be described in part by how many times they proceed around the origin for each oscillation of the forcing function. This measures the output frequency relative to the input frequency (1.0 in this case). The resulting ratio is called the rotation number, and its calculation [6.13] is shown in Figure 6.9.

We see here that there are overlapping sets where two stable periodic solutions having different periods coexist. Since these two solutions lie in the annulus described above, their domains of attraction must be very mixed up [see 6.4]!

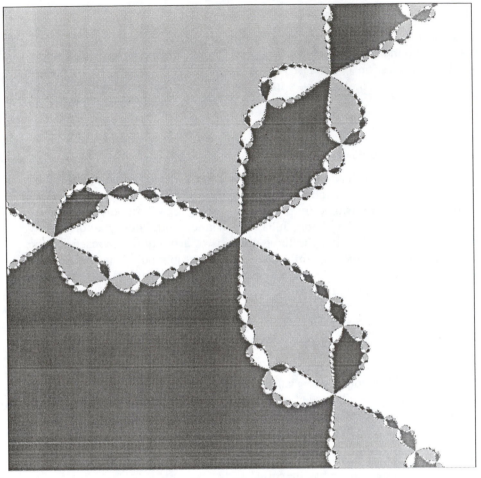

FIGURE 6.7. Julia set.

6.3.3 *Fractal Measures*

These examples show that quite complicated basin boundaries can occur in simple problems. It is therefore of some interest to understand "very crinkly" curves. Figure 6.10 describes the construction of a sequence of curves that have common end points but become successively more complex; in particular, they become longer even though they keep a recognizable shape.

In four steps, the curve becomes quite complicated, and as the process continues, we rapidly lose track of it. We can view each step of this process as a measurement. For example, at first, we use a rod of length 1; second, one of length 1/4, and so on. Obviously, the process results in some limiting set, but the lengths of the approximating curves increase without bound.

We are interested in how the length of the approximating curve changes with the length of the measure used. For example, if $L(\varepsilon)$ denotes the length

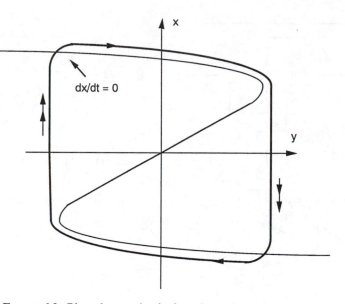

FIGURE 6.8. Phased portrait of a forced van der Pol equation.

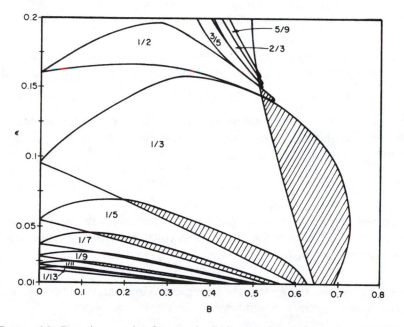

FIGURE 6.9. Rotation number for van der Pol's equation (redrawn from [6.13]).

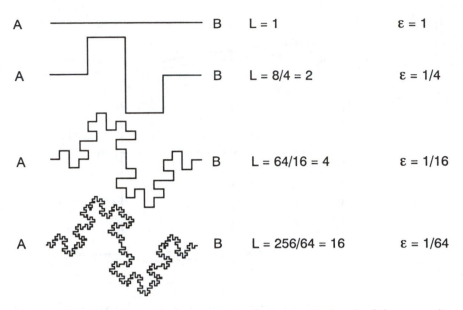

FIGURE 6.10. Complicated basin boundaries. L denotes the length of the curve. At each step, the linear segments are split into four equal parts and rearranged. ε is the length of the measuring rod at each step.

of the approximating line that is based on using a measure ε, then, as for the example in Figure 6.10, the length doubles when the measure is quartered

$$L(\varepsilon/4) = 2L(\varepsilon)$$

This gives a recursion formula between the lengths using different measures. We can solve this recursion using the z transform; that is, we set $L(\varepsilon) = \varepsilon^\alpha$. Solving for α in the result gives

$$\alpha = -\frac{\log 2}{\log 4} = -\frac{1}{2}$$

A trivial example is where we successively partition a straight-line segment. In this case, the length of the line remains constant no matter what value of ε we take. For example,

$$L(\varepsilon/2) = 1$$

for all ε. Thus, $\alpha = 0$ for a straight-line fractal. However, our intuition is that a straight-line segment should have dimension 1, so we define

$$d_c = 1 - \alpha$$

Thus, for the straight-line fractal $d_c = 1$ and for the curve in the Figure 6.10, $d_c = 3/2$. The notation d_c is used to indicate that we have computed the Kolmogorov *capacity of the curve* [6.14]. Figure 6.11 shows some other examples.

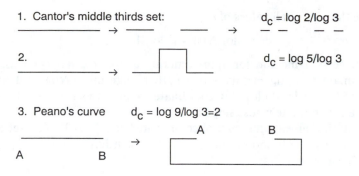

FIGURE 6.11. Capacities of various fractal curves.

6.3.4   *Simulation of Fractal Curves*

Now, suppose that **A** and **B** make up a set whose area is 1. Moreover, suppose that these two sets are separated by a fractal curve having dimension d_c.

We now perform calculation in which we place a point in this set and observe whether it lies in **A** or **B**. We suppose that our placement of the point is accurate only to within some tolerance, say δ. Therefore, our experiment corresponds to placing a disk of radius δ on the diagram. We say that the experiment is undecidable if the disk intersects the boundary of **A** or the boundary of **B**, and so the relevant thing here is the area of a δ-neighborhood of the boundaries of these two sets. This area should be the length of the boundary of **A** plus the length of the boundary of **B** multiplied by δ. However, a fractal boundary has infinite length. Let $L(\delta)$ denote the length of the common boundary of the two sets using a measuring rod of length δ. Then the area covered by this approximation is proportional to

$$L(\delta)\delta = \delta^{1+\alpha} = \delta^{2-d_c}$$

For example, if $\delta = 1.0 \times 10^{-10}$ and $d_c = 1.9$, then the area of this "flattened" boundary is comparable to 0.1. Thus, ten orders of magnitude of precision are lost [6.15]!

Fractals in Higher Dimensions

Kolmogorov's capacity is easy to define for sets in an arbitrary Euclidean space. Consider a bounded set, say S, in E^M. Given a number ε, let $N(\varepsilon)$ denote the number of nonoverlapping (M-dimensional) cubes needed to cover S. Define

$$d_c = \lim_{\varepsilon \to 0^+} \frac{\log N(\varepsilon)}{\log 1/\varepsilon}$$

This limit need not exist, but when it does, this is called the fractal dimension of S.

Note that for small values of ε

$$\log N(\varepsilon) \sim d_c \log(1/\varepsilon)$$

This suggests a method for approximating d_c for a given curve C. Fix ε, approximate C using measures of length ε, determine $N(\varepsilon)$, and plot the values of N and $1/\varepsilon$ on log–log coordinates. Doing this for several values of ε gives a straight line whose slope is d_c.

Hausdorff defined a measure of complicated sets [6.16]. Let a set S in E^2 be covered by a collection of disks, say D_i, which have radii ε_i, respectively. We define a number

$$\Delta(d) = \inf_{\text{partitions}} \sum_i \varepsilon_i^d$$

where the infimum is taken over all possible coverings of S by disks having radii ε_i. Hausdorff showed that $\Delta = \infty$ when d is large, and $\Delta = 0$ when d is small. There is a unique value of d for which $0 < \Delta(d) < \infty$. This value of d, say d_H, is called the *Hausdorff dimension* of S. It is clear that $d_c \geq d_H$ for any set, since d_c is obtained by using a special covering of S. (See Ref. 6.19 for a further discussion of these topics.)

6.4   *Exercises*

6.1. Show that $(Lx, y) = (x, Ly)$ when $Lx = x'' + \mu^2 x$ and $x(t)$ and $y(t)$ are functions whose second derivatives are square integrable and the notation (f, g) is the inner product notation defined in Section 6.1. Compare this result with the analogous calculation for matrices; that is, for which $N \times N$ matrices A is $A\mathbf{x} \cdot \mathbf{y} = \mathbf{x} \cdot A\mathbf{y}$ for all N vectors $\mathbf{x}$ and $\mathbf{y}$?

6.2. Verify that the amplitude A of successive iterates in Duffing's method solves

$$\frac{3\beta A^3}{4} + (\alpha - \mu)A - C = 0$$

as in Section 6.2.2.

6.4. Reproduce Poincare's mappings for Duffing's equation in Figures 6.4, 6.5, and 6.6, using computer simulations.

6.5. Show $d_C \geq d_H$, where these numbers are defined in Section 6.3.4.

7
Methods of Averaging

Regular perturbation methods are based on Taylor's formula and on implicit function theorems. However, there are many problems to which Taylor's formula cannot be applied directly, in which case perturbation methods based on multiple time or space scales can often be used, sometimes even for chaotic systems.

Relatively fast and slow scales can occur in many ways. For example, suppose that A is a 4×4 matrix having eigenvalues

$$\lambda_1 = \rho + i\omega, \qquad \lambda_2 = \rho - i\omega, \qquad \lambda_3 = \mu, \qquad \lambda_4 = \eta$$

where $|\rho| \ll \omega$ and $\eta \ll \mu < 0$. The spectral decomposition of A is (see Section 1.2.1)

$$A = \sum_{j=1}^{4} \lambda_j P_j$$

where P_1, and so on, are projection matrices satisfying the conditions

$$P_j P_k = 0 \qquad \text{if } j \neq k \text{ and } P_k^2 = P_k$$

The solution of the initial value problem

$$\frac{dx}{dt} = Ax, \qquad x(0) \text{ is given}$$

can be derived using this notation:

$$x(t) = \sum_{j=1}^{4} e^{\lambda_j t} P_j x(0)$$

The first two terms in this sum are modulated at a slower rate than they oscillate since the *amplification rate* ρ is smaller than the *frequency* ω. The third term is slowly modulated relative to the fourth, which decays rapidly to zero since $\eta \ll \mu$. If we write $\varepsilon = \rho/\omega$, then $\varepsilon \ll 1$ and the first two terms can be written as

$$e^{\omega(\varepsilon + i)t} P_1 x(0) + e^{\omega(\varepsilon - i)t} P_2 x(0)$$

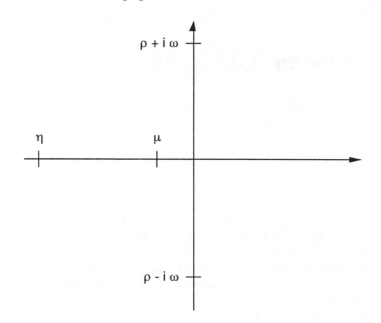

FIGURE 7.1. Spectrum of A.

We could say that these modes vary on two time scales: ωt (oscillation) and a slower one, $\varepsilon \omega t$ (modulation).

Similarly, setting $\varepsilon' = \mu/\eta$ we can write the last two terms as

$$\exp(\mu t) P_3 x(0) + \exp(\mu t/\varepsilon') P_4 x(0)$$

This is a sum of transients that decay on two time scales: μt and a faster one, $\mu t/\varepsilon'$.

Plotting the eigenvalues of A can suggest what scales govern the solution's behavior, as shown in Figure 7.1. Eigenvalues lying near the imaginary axis give rise to *slowly modulated oscillations*; eigenvalues lying near the origin correspond to *quasistatic modes*; and other eigenvalues lying near the real axis describe *rapidly growing* or *decaying modes*.

This example shows that small parameters can appear in linear systems as ratios of eigenvalues, and so the system's solution can include various scales. There are important implications of this for numerical methods since stability and accuracy of numerical schemes depend on step sizes being appropriate to changes in solutions. Although eigenvalues can be used to discover various scales in linear problems, there is no reliable method for finding their analogs in nonlinear problems. In fact, nonlinear problems typically involve scales that change with the system's present state. Some aspects of this are studied in Chapter 8.

In this chapter we study highly oscillatory problems, and in the next we study problems with rapid transients. Highly oscillatory systems are studied using averaging methods; we first state and prove a general averaging result that is used throughout this chapter. The ideas developed here are similar to those of stability under persistent disturbances (Chapter 3), but in addition they provide methods for constructing precise approximations to solutions. A useful approximation shortcut that is based on a multiscale expansion formalism is also developed in Section 7.1.

The general theory is later specialized and applied to a variety of problems. First, in Section 7.2 linear problems are studied in detail, especially certain linear feedback problems where detailed computations can be carried out. In Section 7.3, the multiscale methodology is revised and applied to nearly linear difference equations. We study a variety of iterations using those results.

Weakly nonlinear problems are studied in Section 7.4; in particular, we revisit van der Pol's and Duffing's equations as well as more general systems of weakly coupled harmonic oscillators. Special cases of these are problems for (angular) phase variables. In Section 7.5, three separate approaches to phase-amplitude systems are derived: the rotation vector method (RVM), Bogoliuboff's near-identity transformation for differential equations on a torus, and the Kolmogoroff-Arnol'd-Moser (KAM) theory for quasiperiodic solutions. Section 7.6 introduces how microscopic spatial structure can be described and used to derive macroscopic properties of fine periodic structures in space. This is the method of homogenization. Finally, computational methods that are based on averaging are described in Section 7.7.

Averaging nonlinear conservative systems can be delicate because small divisors can occur. These arise through nonlinear feedbacks that create internal resonance. Many important techniques have been developed to deal with small divisors, culminating (to date) in the RVM and the KAM theories. These theories show how to deal with oscillatory problems in which there is no dissipation. When dissipation is present in a system, the theorem on nonlinear averaging for mean-stable systems is used to construct solutions.

Problems encountered in highly oscillatory system are already apparent in linear systems. The following example is quite useful. Consider the differential equation

$$\frac{dx}{dt} = U^{-1}(t)BU(t)x$$

where x is in E^2, U is the rotation matrix

$$U(t) = \begin{pmatrix} \cos \omega t & -\sin \omega t \\ \sin \omega t & \cos \omega t \end{pmatrix}$$

and

$$B = \begin{pmatrix} a & b \\ c & d \end{pmatrix}$$

This system was studied in Section 1.4. The change of variables $y = Ux$ takes this system into

$$\frac{dy}{dt} = (B + \omega J)y$$

where J is Jacobi's matrix

$$J = \begin{pmatrix} 0 & -1 \\ 1 & 0 \end{pmatrix}$$

U is a matrix that rotates a vector through an angle ωt, so we refer to the system for x as being the result of spinning the system for y. The eigenvalues of this system are given by the solutions of the equation

$$\lambda^2 - (a + d)\lambda + ad - bc + \omega(c - b) + \omega^2 = 0$$

We consider three interesting examples:

Example 1. Spinning can destabilize a linear system. Let

$$B = \begin{pmatrix} -1 & b \\ 0 & -2 \end{pmatrix}$$

Then the eigenvalues of $U^{-1}AU$ are -1 and -2, but the eigenvalues of $B + \omega J$ are

$$\lambda = \frac{-3 \pm \sqrt{1 + 4b\omega - 4\omega^2}}{2}$$

The neutral stability condition for this system (viz., $\text{Re }\lambda = 0$) occurs when the real part of the eigenvalues are zero: This happens when $4b\omega - 4\omega^2 = 8$. Therefore, there are two regions, **S** and **U**, in the plane such that if (ω, b) is in **S**, then $y = 0$ is a stable spiral node. If (ω, b) is in **U**, then it is an unstable node! Thus, spinning the system (ω large) can destabilize it. Coppel showed that in general slow oscillations ($|\omega| \ll 1$) cannot destabilize the system if the eigenvalues of A have negative real parts [7.1].

Example 2. Spinning can stabilize a hyperbolic system. Let

$$B = \begin{pmatrix} 0 & 1 \\ 1 & 0 \end{pmatrix}$$

The eigenvalues of this matrix are ± 1, but the eigenvalues of $B + \omega J$ are $\pm \sqrt{(1 - \omega^2)}$. Therefore, if $|\omega| < 1$, $x = 0$ is a hyperbolic point (i.e., it is conditionally stable), but, if $\omega^2 > 1$, it is a stable node. Thus, spinning can stabilize an unstable node.

Finally, let's expand $U^{-1}(t)BU(t)$ in its Fourier series:

$$U^{-1}(t)BU(t) = B_0 + \text{h.f.t.}$$

where h.f.t. denotes higher frequency terms. In this case, the matrix B_0 is

$$B_0 = \begin{bmatrix} \dfrac{a+d}{2} & -\dfrac{c-b}{2} \\ \dfrac{c-b}{2} & \dfrac{a+d}{2} \end{bmatrix}$$

whose eigenvalues are $(a + d \pm (c - b)i)/2$.

The next example shows that the average B_0 and the Floquet matrix of $B + \omega J$ can coincide.

Example 3. Floquet matrix for the system. Our example system is periodic in t, and so Floquet's theory, which shows that such a system can be transformed into a system with constant coefficients, leads to the matrix R of Section 1.4; in this case, $R = B + \omega J$. The eigenvalues of $B + \omega J$ for large ω are near those of $B_0 \pm i\omega$, up to order $O(1/\omega^2)$! They are given by

$$\lambda = \pm i\omega + \frac{(a+d) \pm i(c-b)}{2} + O\left(\frac{1}{\omega}\right)$$

Thus, the eigenvalues of the average $B_0 \pm i\omega$ and of $B + \omega J$ are nearly identical when ω is large. Recall that Floquet's Theorem states that a fundamental matrix for the system for x is $P(t) \exp(Rt)$. In this case it is

$$U(t) \exp((B + \omega J)t)$$

which is approximately $\exp(B_0 t)$.

In summary, we have seen in these examples that spinning can destabilize a stable system, and it can stabilize an unstable system. Finally, we have seen that the average of a highly oscillatory system is closely related (by Floquet's theory) to the full problem. We investigate this connection next in a more general setting.

7.1   Averaging Nonlinear Systems

Consider the system of differential equations

$$\frac{dx}{dt} = \varepsilon f(t, \varepsilon t, x, \varepsilon)$$

or its equivalent,

$$\frac{dx}{d\tau} = f\left(\frac{\tau}{\varepsilon}, \tau, x, \varepsilon\right)$$

where $x, f \in E^n$, ε is a small real parameter, say $|\varepsilon| \ll 1$, and the slow time variable $\tau = \varepsilon t$ is restricted to some finite interval $0 \le \varepsilon t \le T < \infty$. Suppose the following hypothesis is true.

Hypothesis H1. $f(t, \tau, x, \varepsilon)$ is a smooth function of its arguments for $0 \leq t \leq T/\varepsilon$, $0 \leq \tau \leq T$, for x in some domain G lying in E^n, and for ε near zero. Moreover, suppose that f is an almost periodic function of t, uniformly in the other variables. Specifically, we assume that f can be expanded in a uniformly convergent generalized Fourier series

$$f(t, \tau, x, \varepsilon) = f_0(\tau, x, \varepsilon) + \sum_{n=1}^{\infty} f_n(\tau, x, \varepsilon) e^{i\omega_n t}$$

where the *frequencies* ω_n satisfy $\omega_n \neq 0$ for $n \geq 1$.

The smooth functions f_n are the *amplitudes* of the various modes of rapid oscillation of f relative to t. When convenient, we write $\omega_0 = 0$. Functions of this kind can be indistinguishable from chaotic ones in practice, but there are fine points of distinction between chaotic functions, say ones that have a power spectrum supported by an interval of frequencies, and functions that have a convergent generalized Fourier expansion. In particular, we can average the latter functions to good use.

If we attempt to average f with respect to t, we must evaluate the limit

$$\langle f \rangle(\tau, x, \varepsilon) \equiv \lim_{T \to \infty} \frac{1}{T} \int_0^T \sum_{n=0}^{\infty} f_n e^{i\omega_n t} \, dt = f_0(\tau, x, \varepsilon)$$

Thus, f_0 is the mean value of f with respect to t provided the series of integrals converges. Unfortunately, the series appearing here need not converge. For example, if $f_n = n^{-2}$ and $\omega_n = 2^{-n}$, then, although the series can be integrated term-wise, the small numbers ω_n appear in the denominators of the integral, and the resulting series does not converge. This illustrates the small divisor problem that is discussed in Section 7.5.

7.1.1   The Nonlinear Averaging Theorem

The next condition ensures that the average of f exists.

Hypothesis H2. Let $g = f - f_0$, and suppose that the integral

$$\int_0^t g(t', \tau, x, 0) \, dt' = \int_0^t [f(t', \tau, x, 0) - \langle f \rangle(\tau, x, 0)] \, dt'$$

is bounded uniformly for $0 \leq t \leq T/\varepsilon$ (and so for $0 \leq \tau \leq T$) and for $x \in G$.

This condition is not the "best possible" one, but it suffices for our purposes here. Condition H2 is usually satisfied, although it can be difficult to verify in applications. However, if f is a periodic function of t or if it is a finite trigonometric polynomial in t, then this condition is satisfied. On the other hand, as we have seen, if the Fourier series for f is sufficiently lacunary, say $\omega_n = 2^{-n}$, then this condition might not be satisfied. More general conditions for averaging nonlinear problems are given in Refs. 7.2 and 7.3.

We have the following theorem

Nonlinear Averaging Theorem. *Suppose that conditions H1 and H2 are satisfied and suppose that the averaged system*

$$\frac{dX}{d\tau} = f_0(\tau, X, 0), \qquad X(0) = \eta \tag{7.1}$$

has a unique solution lying in G for $0 \le \tau \le T$. If ε_0 is sufficiently small and if $|\varepsilon| \le \varepsilon_0$, then:

1. *There is a unique solution of the problem*

$$\frac{dx}{dt} = \varepsilon f(t, \varepsilon t, x, \varepsilon), \quad x(0) = \eta$$

   for $0 \le t \le T/\varepsilon$.
2. *The solution lies in G.*
3. *There is a constant K depending on T and ε_0 such that*

$$|x(t) - X(\varepsilon t)| \le K|\varepsilon|$$

   for $0 \le t \le T/\varepsilon$.

This result shows that the solution of the averaged equation (7.1) approximates the solution of the full problem having the same initial data over large (growing) intervals, $0 \le t \le T/\varepsilon$.

Proof of the Nonlinear Averaging Theorem. The theorem is proved in a more general setting in Ref. 7.3. The proof here follows from a straightforward application of Gronwall's inequality.

Let $z = x(t, \varepsilon) - X(\varepsilon t)$. This function satisfies the initial value problem

$$\frac{dz}{dt} = \varepsilon f(t, \varepsilon t, z + X, \varepsilon) - \varepsilon f_0(\varepsilon t, X, 0)$$

$$= \varepsilon[f_x(t, \varepsilon t, X, 0)z + f(t, \varepsilon t, X, 0) - f_0(\varepsilon t, X, 0) + O(\varepsilon) + O(|z|^2)]$$

$$z(0) = 0$$

where f_x denotes the Jacobian matrix of the components of f with respect to the components of x.

First, we define the linear problem's fundamental solution by $U(t, s)$ which can be determined from the differential equation

$$\frac{dU}{dt} = \varepsilon f_x(t, \varepsilon t, X(\varepsilon t), 0)U, \qquad U(s, s) = I$$

Using this, we can convert the initial value problem for x into an equivalent integral equation for z:

$$z(t) = \varepsilon \int_0^t U(t,s)(f(s, \varepsilon s, X(\varepsilon s), 0) - f_0(\varepsilon s, X(\varepsilon s), 0) + O(\varepsilon) + O(|z|^2)) \, ds$$

Let us define $g(s) = f(s, \varepsilon s, X(\varepsilon s), 0) - f_0(\varepsilon s, X(\varepsilon s), 0)$, and integrate this formula once by parts. The result is

$$\varepsilon \int_0^t U(t,s) g(s) \, ds = \varepsilon U(t,s) \int_0^s g(s') \, ds' \Big|_0^t - \varepsilon \int_0^t \frac{dU}{dt} \int_0^s g(s') \, ds'$$

This shows that

$$\varepsilon \int_0^t U(t,s) g(s) \, ds = \varepsilon \int_0^t g(s) \, ds - \varepsilon^2 \int_0^t f_x U(t,s) \int_0^s g(s') \, ds'$$

It follows from condition H2 that, for $0 \le t \le T/\varepsilon$, this integral is bounded by $K_1 \varepsilon$ where K_1 is a constant that depends on ε_0 and T. Therefore,

$$|z(t)| \le \varepsilon \int_0^t |U(t,s)| O(|z|^2) \, ds$$

It follows from Gronwall's inequality (see Chapter 3) that $|z(t)| = O(\varepsilon)$ uniformly for $0 \le t \le T/\varepsilon$. ∎

7.1.2 Averaging Theorem for Mean-Stable Systems

Now suppose that $T = \infty$ in conditions H1 and H2. Adding a stability condition can result in an approximation that is valid on the entire half-line $0 \le t < \infty$. We say that the system is *mean stable* if condition H3 is true.

Hypothesis H3. The averaged equation

$$\frac{dX}{d\tau} = f_0(\tau, X, 0)$$

has a rest point, say X^*, that is exponentially asymptotically stable.

With this additional condition, we have the following theorem.

Averaging Theorem for Mean-Stable Systems. *Let f satisfy conditions H1, H2, and H3. Let X denote the solution of the initial value problem*

$$\frac{dX}{d\tau} = f_0(\tau, X, 0), \qquad X(0) = \eta$$

for $0 \le \tau < \infty$. If η is near X^* and if $\varepsilon > 0$ is sufficiently small, then the problem

$$\frac{dx}{dt} = \varepsilon f(t, \varepsilon t, x, \varepsilon), \qquad x(0) = \eta$$

has a unique solution for $0 \le t < \infty$, and

$$x(t, \varepsilon) = X(\varepsilon t) + O(\varepsilon)$$

where the error estimate here holds as $\varepsilon \to O^+$ uniformly for $0 \le t < \infty$.

This result is quite strong since it gives a valid approximation of the system's solution over the entire half-line $t \ge 0$. The stability condition H3 required for the result is quite restrictive, but it can be relaxed somewhat to stability under persistent disturbances or uniform asymptotic stability [7.4]. Note that the sign of ε must be positive in this result to ensure that the stability estimates hold for $t \to \infty$. This result also highlights the duality between $\tau \to \infty$ and $\varepsilon \to 0$: The solution can be written as $x(\tau/\varepsilon, \varepsilon) = X(\tau) + O(\varepsilon)$, which decomposes x into two components, the behavior as $\tau \to \infty$ and the behavior as $\varepsilon \to 0$.

Proof of the Mean-Stable Averaging Theorem. The proof of this result proceeds exactly like that of the last section's except that we must extend the estimate to the entire line $t \ge 0$. First, note that condition H3 ensures that the fundamental solution $U(t, s)$ satisfies an estimate of the form

$$|U(t, s)| \le K e^{-\alpha(t-s)}$$

for $0 \le s \le t < \infty$ and for some positive constants K and α. Then, we define $u(t) = |z(t)| e^{\alpha t}$ as we did in using Gronwall's Lemma in proving the linear stability theorem in Section 3.2. It follows that $u(t)$ increases like a slow exponential function if ε is sufficiently small. Therefore, $|z(t)|$ decays at an exponential rate with amplitude at most $O(\varepsilon)$. ∎

7.1.3 A Two-Time-Scale Method for the Full Problem

Based on the last two theorems, we observe that two time scales (t and εt) appear in the solutions, and it is useful to exploit this fact. To do this, we introduce as a solution of this equation a function

$$x = u(t, \tau, \varepsilon)$$

where t and $\tau = \varepsilon t$ are treated as independent variables (when convenient). Keep in mind that this guess for the form of a solution must be validated *a posteriori*. We suppose that u is a smooth function of ε, so that

$$u(t, \tau, \varepsilon) = u_0(t, \tau) + \varepsilon u_1(t, \tau) + \varepsilon^2 u_2(t, \tau) + \cdots$$

where the coefficients are assumed to be bounded functions for $0 \le t, \tau < \infty$ to ensure that this expansion makes sense as $\varepsilon \to 0$. The problem becomes

$$\frac{\partial u}{\partial t} + \varepsilon \frac{\partial u}{\partial \tau} = \varepsilon f(t, \tau, u, \varepsilon)$$

We hope that the dependence of x on the two time scales t and τ is correctly accounted for by this, and we proceed to construct Taylor's expansion for u by successively differentiating the equation with respect to ε:

$$\frac{\partial u_0}{\partial t} = 0$$

$$\frac{\partial u_1}{\partial t} = f(t, \tau, u_0, 0) - \frac{\partial u_0}{\partial \tau}, \cdots$$

The first equation implies that u_0 does not depend on t, so we write

$$u_0 = U_0(\tau).$$

The second equation can be integrated and its solution is

$$u_1(t, \tau) = U_1(\tau) + \int_0^t f(t', \tau, U_0(\tau), 0) \, dt' - t \frac{\partial U_0}{\partial \tau}$$

where U_1 is an unknown function. Since u_1 is to be bounded, we must have that

$$\frac{dU_0}{d\tau} = \lim_{T \to \infty} \frac{1}{T} \int_0^T f(t', \tau, U_0(\tau), 0) \, dt' = f_0(\tau, U_0(\tau), 0)$$

The process can be continued to determine more coefficients in the expansion of u. At least we see that $U_0(\tau) = X(\tau)$ as we determined in the nonlinear averaging theorem.

It is perhaps questionable that we can average over the t variable while holding the τ variable fixed. Still, the results of this nonrigorous calculation agree with those in the averaging theorems just proved, and so they produce the correct equations when the conditions of those theorems are satisfied.

The two-time-scale scheme provides a good probe to use in studying novel problems, as we see in studies of difference equations in Section 7.3 and in homogenization problems in Section 7.6.

7.2   Highly Oscillatory Linear Systems

It is useful to go through the steps of the two averaging theorems for linear systems.

7.2.1 $dx/dt = \varepsilon B(t)x$

Systems of the form

$$\frac{dx}{dt} = \varepsilon B(t)x, \qquad x(0, \varepsilon) \text{ given}$$

where ε is a small parameter and $B(t)$ is a matrix of smooth quasiperiodic functions, say ones having uniformly convergent generalized Fourier expansions, are difficult to solve in general. Even if B is a matrix of constants, calculating values of the exponential $e^{\varepsilon B t}$ can pose serious computational problems [see 7.5], and if $B(t)$ is periodic, determination of Floquet's matrix is usually difficult. Still, systems in this form are quite useful, and we study next how they can be analyzed using the method of averaging.

Let us first rewrite the equation as

$$\frac{dx}{dt} = \varepsilon B_0 x + \varepsilon(B(t) - B_0)x$$

where B_0 denotes the average of $B(t)$:

$$B_0 = \lim_{T \to \infty} \frac{1}{T} \int_0^T B(t')\,dt'$$

Thus, B_0 is made up of all of the constant terms in $B(t)$. Note that if $B(t)$ is a matrix of functions that are 2π periodic in t, then

$$B_0 = \frac{1}{2\pi} \int_0^{2\pi} B(t)\,dt$$

The averaged equation case is

$$\frac{dX}{dt} = \varepsilon B_0 X$$

and its solution is

$$X(\varepsilon t) = e^{\varepsilon B_0 t} X_0$$

There are three cases depending on what interval we are interested in approximating over. These are summarized in the following theorem:

Averaging Theorem for Linear Systems.

1. $0 \le t \le T\,(< \infty)$. *On this interval,*

$$x(t, \varepsilon) = x_0 + O(\varepsilon)$$

where the error estimates hold uniformly for t restricted to intervals that remain bounded as $\varepsilon \to 0$.

2. $0 \le t \le T/\varepsilon$. *In this case,*

$$x(t, \varepsilon) = \exp(\varepsilon B_0 t)x_0 + O(\varepsilon)$$

where the error estimate holds uniformly for $0 \le t \le T/\varepsilon$. Therefore, we have that

$$x(t, \varepsilon) = X(t\varepsilon) + O(\varepsilon)$$

3. $0 \le t < \infty$. *Suppose that there are positive constants K and α such that*

$$|e^{\varepsilon B_0 t}| \le K e^{-\alpha \varepsilon t}$$

for $0 \le t < \infty$. Then $x(t, \varepsilon) = X(t\varepsilon) + O(\varepsilon)$ where the error estimate holds as $\varepsilon \to 0^+$ uniformly in $t, 0 \le t < \infty$. For example, if all eigenvalues of B_0 have *negative real parts, then the averaged equation gives an approximation to x that is useful for all t.*

These results follow from direct applications of the nonlinear averaging theorems to the linear problem. The first of these three cases is not very interesting since no influence of B is felt to the order of accuracy $O(\varepsilon)$. However, the other two cases are of great interest.

7.2.2   *Linear Feedback System*

A linear feedback system of the form

$$\frac{dx}{dt} = (A + \varepsilon C)x$$

where A is an oscillatory matrix (i.e., A is a diagonalizable matrix having purely imaginary eigenvalues) can be analyzed using the averaging theorems. We spin the system as before. Let

$$x = \exp(At)y$$

Then

$$\frac{dy}{dt} = \varepsilon e^{-At}Ce^{At}y$$

where now the coefficient matrix is a linear combination of oscillatory functions that can be found using the spectral decomposition of A.

In fact, if A has the form

$$A = \sum_{j=1}^{M} i\omega_j P_j$$

we can calculate B_0 directly. Since

$$B(t) = e^{-At}Ce^{At}$$

is a matrix of trigonometric polynomials, we know that B_0 exists, and it is given by the formula

$$B_0 = \sum_{\omega_j = \omega_k} P_j C P_k$$

where the sum is taken over all indices j and k for which $\omega_j = \omega_k$.

7.2.3   *Connections Between Averaging and Laplace's Method*

Laplace studied highly oscillatory integrals, and he derived various asymptotic expansions for them. A particular example was described in Section 5.1.4. In general those methods are based on the method of steepest descent, but they take a simpler form in the case we considered in the last section where we studied the equation

$$\frac{dy}{d\tau} = e^{-A\tau/\varepsilon}Ce^{A\tau/\varepsilon}y$$

Integrating this equation gives an equivalent integral equation for $y(\tau)$:

$$y(\tau) = y(0) + \int_0^\tau e^{-As/\varepsilon}Ce^{As/\varepsilon}y(s)\,ds$$

$$= y(0) + B_0 \int_0^\tau y(s)\,ds + \int_0^\tau (e^{-As/\varepsilon}Ce^{As/\varepsilon} - B_0)y(s)\,ds$$

Integrating the last integral once by parts and using the conditions of the previous section shows that it is $O(\varepsilon)$. In fact, Laplace's method reduces to the Riemann-Lebesgue Lemma in this case since the only terms surviving in the last integral are those that are oscillatory functions of s/ε multiplied by $y(s)$. The Riemann-Lebesgue Lemma shows that such integrals approach zero as $\varepsilon \to 0$.

7.3   Averaging Rapidly Oscillating Difference Equations

The two-time-scale formalism can be very useful in studying problems that fall outside the usual differential equations considered up to this point. For example, consider the difference equation

$$x_{n+1} = x_n + \varepsilon f(n, x_n, \varepsilon)$$

where $x, f \in E^N$. Given x_0, this recursion defines a unique sequence $\{x_n\}$, which we refer to as being a solution of the recursion. The question asked here is: Can we construct an approximate solution that is valid when $|\varepsilon| \ll 1$?

We can begin to answer this question by using multiple time scales. We observe that if f is a function of x only, then this recursion is trying hard to be a differential equation. In fact, if $X(s)$ is a smooth function such that $X(n\varepsilon) = x_n$, then the recursion defines a forward Euler scheme for the differential equation

$$\frac{dX}{ds} = f(X)$$

Motivated by this, we make a guess for the solution as being a function of the two time scales n and εn:

$$x_n = u(n, \varepsilon n, \varepsilon) = u_0(n, \varepsilon n) + \varepsilon u_1(n, \varepsilon n) + \varepsilon^2 u_2(n, \varepsilon n) + \cdots$$

We write $s = \varepsilon n$, and in terms of u, the recursion takes the form

$$u(n + 1, s + \varepsilon, \varepsilon) = u(n, s, \varepsilon) + \varepsilon f(n, u, \varepsilon)$$

Setting $\varepsilon = 0$ in this equation gives

$$u_0(n+1, s) = u_0(n, s)$$

Differentiating the equation with respect to ε and setting $\varepsilon = 0$ in the result gives

$$u_1(n+1, s) = u_1(n, s) + f(n, u_0(n, s), 0) - \frac{\partial u_0}{\partial s}$$

and so on.

The first equation shows that u_0 does not depend on n, and so we write

$$u_0(n, s) = U_0(s)$$

The second equation shows that

$$u_1(n, s) = U_1(s) + \sum_{j=0}^{n-1} f(j, U_0(s), 0) - n \frac{dU_0}{ds}$$

If u_1 remains bounded as $n \to \infty$, then dividing both sides of this equation by n and passing to the limit $n = \infty$ gives

$$\frac{dU_0}{ds} = \lim_{n \to \infty} \frac{1}{n} \sum_{j=0}^{n-1} f(j, U_0(s), 0)$$

We take for initial data here the value

$$U_0(0) = x_0$$

Hypothesis H4. Suppose that the average

$$f_0(U) = \lim_{N \to \infty} \frac{1}{N} \sum_{j=0}^{N-1} f(j, U, 0)$$

exists and defines a smooth function of U. Moreover, suppose that the difference

$$\sum_{j=0}^{N-1} [f(j, U, 0) - f_0(U)]$$

remains bounded uniformly in N and U.

Difference Equation Averaging Theorem. *Suppose that condition H4 is satisfied for the recursion*

$$x_{n+1} = x_n + \varepsilon f(n, x_n, \varepsilon)$$

and that each function $f(n, x, \varepsilon)$ is a smooth function of its arguments x (in E^N) and ε (near zero) for $n = 0, 1, 2, \cdots$. Let $U(s)$ be determined from the differential equation

$$\frac{dU}{ds} = f_0(U), \qquad U(0) = x_0$$

Then

$$x_n = U(\varepsilon n) + O(\varepsilon)$$

where the error estimate holds (as $\varepsilon \to 0$) uniformly for $n = 0, 1, \cdots, N[1/\varepsilon]$. Here N is any fixed positive integer and $[1/\varepsilon]$ denotes the integer part of $1/\varepsilon$, that is, the largest integer less or equal to $1/\varepsilon$.

Proof of the Difference Equation Averaging Theorem. The proof is accomplished by successfully estimating the difference

$$\delta_n = x_n - U(\varepsilon n)$$

We have

$$\delta_{n+1} = x_{n+1} - U(\varepsilon n + \varepsilon)$$
$$= x_n + \varepsilon f(n, x_n, \varepsilon) - U(\varepsilon n) - \varepsilon f_0(U(\varepsilon n), 0) + O(\varepsilon^2)$$
$$= \delta_n + \varepsilon(f(n, \delta_n + U(\varepsilon n), \varepsilon) - f_0(U(\varepsilon n))) + O(\varepsilon^2) + \varepsilon O(|\delta_n|^2)$$

It follows that

$$\delta_n = O(\varepsilon) + \varepsilon \sum_{k=0}^{n-1} O(|\delta_k|^2)$$

There is a constant K such that $|O(\varepsilon)| \le K\varepsilon$ and $O(|\delta_k|^2) \le K|\delta_k|^2$ for $k = 0, 1, \cdots, n-1$. Now suppose that $|\delta_k| \le 2K\varepsilon$ for $k \le n-1$. Then

$$|\delta_n| \le K(\varepsilon + 4NK^2\varepsilon^2) \le 2K\varepsilon, \qquad \text{if } 4K^2N\varepsilon < 1$$

Therefore, if $\varepsilon < 1/(4K^2N)$, by mathematical induction we have that $\delta_n = O(\varepsilon)$ uniformly for $n = 1, \cdots, N[1/\varepsilon]$ [see 7.6]. ∎

If U approaches a rest point of the differential equation, say $U \to U^*$, where $f_0(U^*) = 0$ and U^* is exponentially asymptotically stable, then the results of the averaging theorem hold uniformly for $0 \le n < \infty$.

7.3.1   Linear Difference Schemes

As we did for linear differential equations, we can carry out several computations in detail for linear recursions. We consider the problem

$$x_{n+1} = (A + \varepsilon B)x_n, \qquad x_0 \text{ given}$$

Writing $x_n = A^n u_n$ gives

$$u_{n+1} = u_n + \varepsilon A^{-n-1} B A^n u_n$$

Applying the analysis of the preceding section, we might hope to find that

$$u_n = U(n\varepsilon) + O(\varepsilon)$$

where $U(s)$ satisfies the equation

$$\frac{dU}{ds} = \lim_{n\to\infty} \frac{1}{n} \sum_{j=0}^{n-1} A^{-j-1}BA^jU$$

If the coefficient matrix has all of its eigenvalues with negative real parts, then we might hope that the error estimate $O(\varepsilon)$ would hold uniformly for all n.

Therefore, it is of interest to determine the structure of the matrix

$$B_0 = \lim_{N\to\infty} \frac{1}{N} \sum_{n=0}^{N-1} A^{-n-1}BA^n$$

if this limit exists. In order to do this, we assume that A is a discrete oscillatory matrix.

Definition. A is a *discrete oscillatory matrix* if it has a spectral decmposition of the form

$$A = \sum_{j=0}^{N} \lambda_j P_j$$

where the eigenvalues λ_j lie *on* the unit circle in the complex plane ($|\lambda_j| = 1$). Therefore, $\lambda_j = \exp(i\omega_j)$ for some real numbers ω_j. A is a *discrete stable matrix* if all eigenvalues λ_j satisfy

$$|\lambda_j| \leq 1$$

for $j = 1, \cdots, N$, and it is a *discrete asymptotically stable matrix* if

$$|\lambda_j| < 1$$

for $j = 1, \cdots, N$.

We can calculate B_0 directly when A is a discrete oscillatory matrix. In fact, then

$$A^{-n-1}BA^n = \sum_{j=1}^{N} \sum_{k=1}^{N} \lambda_k^n \lambda_j^{-n-1} P_j B P_k$$

Moreover, the geometric sum

$$\frac{1}{n} \sum_{m=0}^{n-1} \lambda_k^m \lambda_j^{-m-1} = \frac{\lambda_j^{-1}}{n} \frac{(\lambda_k/\lambda_j)^n - 1}{(\lambda_k/\lambda_j) - 1}$$

Since the eigenvalues λ lie on the unit circle in the complex plane, this expression approaches zero as $n \to \infty$ unless $\lambda_j = \lambda_k$ in which case it has the value $1/\lambda_j$.

Therefore, we have

$$B_0 = A^{-1} \sum_{\lambda_j = \lambda_k} P_j B P_k$$

where the sum is taken over all indices j and k for which $\lambda_j = \lambda_k$. Thus, B_0 contains at least the terms $P_j B P_j$. The average of B also exists when the eigenvalues of A are arbitrary as long as $P_j B P_k = 0$ whenever $|\lambda_k/\lambda_j| > 1$. The next theorem follows from the Difference Equation Averaging Theorem.

Averaging Theorem for Linear Difference Equations. *Let A be a discrete oscillatory matrix, and let $\{x_n\}$ be a sequence determined by the recursion*

$$x_{n+1} = (A + \varepsilon B)x_n$$

Then for any fixed number M,

$$x_n = A^n \exp(\varepsilon B_0 n)x_0 + O(\varepsilon)$$

where the error estimate holds as $\varepsilon \to 0$ uniformly for $n = 1, \cdots, M$ $[1/\varepsilon]$.

Thus, we can approximate the solution of the recursion using the fundamental solution of the differential equation

$$\frac{dU}{ds} = B_0 U, \qquad U(0) = I$$

Example. A Discrete Rotational System. We consider a case where $N = 2$ and $A = J$, that is, Jacobi's matrix

$$J = \begin{pmatrix} 0 & -1 \\ 1 & 0 \end{pmatrix}$$

This matrix has spectral decomposition

$$J = \frac{i}{2}\begin{pmatrix} 1 & i \\ -i & 1 \end{pmatrix} - \frac{i}{2}\begin{pmatrix} 1 & -i \\ i & 1 \end{pmatrix} \equiv iP_1 + (-i)P_2$$

Now, consider the difference equation

$$x_{n+1} = (J + \varepsilon B)x_n$$

where B is an arbitrary 2×2 matrix.

From the calculations above, we see that the relevant average is given by the formula

$$B_0 = -iP_1 B P_1 + iP_2 B P_2$$

The result of the averaging theorem for linear difference equations shows that

$$x_n = J^n \exp(\varepsilon n B_0) + O(\varepsilon)$$

for $n = 1, \cdots, M$ $[1/\varepsilon]$ for any fixed number M.

Example. Perturbations of a Cyclic Markov Chain. The next example shows how the two-time-scale method can be used to study a random process. A Markov chain describes the dynamics of a system that can move among a finite set of states, say $S_1, \cdots, S_N$, in a random way [7.7]. Let $p_{j,n}$ be the probability that the system is in state S_j at time n and let $P_{j,k}$ denote the probability that the system moves from state S_j to state S_k in one time step. We suppose that these transition probabilities (P_{jk}) do not change with time. As a result, the probability of state j at the sampling time $n + 1$ is the sum of all possible states at the previous step weighted by the probability of transition to state j in one time step. We write this as

$$p_{j,n+1} = \sum_{k=1}^{N} p_{k,n} P_{k,j}$$

for $j = 1, \cdots, N$. The notation can be simplified by introducing the probability density (row) vector

$$\mathbf{p}_n = (p_{1,n}, \cdots, p_{N,n})$$

and the transition probability matrix

$$\mathbf{P} = (P_{j,k})$$

With this, the model becomes

$$\mathbf{p}_{n+1} = \mathbf{p}_n \mathbf{P}$$

The sequence of probability vectors $\{\mathbf{p}_n\}$ describes the evolution of the process.

Two interesting examples of perturbations of cyclic chains are

$$\mathbf{p}_{n+1} = \mathbf{p}_n(I + \varepsilon B)$$

where I is the identity matrix in E^N and

$$\mathbf{p}_{n+1} = \mathbf{p}_n(C + \varepsilon B)$$

where C is a cyclic matrix in E^N:

$$C = \begin{bmatrix} 0 & 1 & & \cdots & & 0 & 0 \\ 0 & 0 & & & & 0 & 0 \\ \vdots & & & \ddots & & & \vdots \\ 0 & 0 & & & & 0 & 1 \\ 1 & 0 & & \cdots & & 0 & 0 \end{bmatrix}$$

We suppose in each case that the row sums of B are all equal to zero and that the matrices in each case ($I + \varepsilon B$ and $C + \varepsilon B$) have nonnegative entries. Therefore, they are stochastic matrices. In the first case, the chain is trivial when $\varepsilon = 0$, and so any distribution will persist in that case. However, when $\varepsilon > 0$, we have from the averaging result that

$$\mathbf{p}_n = U_0(\varepsilon n) + O(\varepsilon)$$

where

$$\frac{dU_0}{ds} = U_0 B$$

Since $I + \varepsilon B$ is a stochastic matrix, all of B's off-diagonal elements must be nonnegative. Thus, $B = D + N$ where D is a diagonal matrix and N is a matrix on nonnegative elements. It follows that

$$U_0 = p_0 \exp(\varepsilon Ds) V_0$$

where V_0 has only nonnegative elements.

Since the eigenvalues of a stochastic matrix all lie in or on the unit circle in the complex plane, the eigenvalues of B must lie at the origin or in the left half-plane. If B has a single eigenvalue at $\lambda = 0$, then all modes of U_0 are damped except for the uniform mode. Thus, if $\varepsilon > 0$, the probability distribution becomes spread out over all possible states.

The second example describes a chain that moves contents regularly from one state to the next and from state S_N to state S_1 when $\varepsilon = 0$. If B has a single eigenvalue at $\lambda = 0$, then the "merry-go-round" chain changes into one that results in a uniform distribution of the process over the states.

Finally, it is useful to consider a nonstationary Markov chain:

$$p_{n+1} = p_n(I + \varepsilon B_n)$$

where B_n is a periodic sequence of matrices, say with period N, for which $I + \varepsilon B_n$ is a stochastic matrix for each $n = 0, \cdots, N - 1$. Such chains arise as models in chronobiology where n denotes weeks and $N = 52$ is a year, or n is in minutes and $N = 1440$ is a day. In this case,

$$p_n = U_0(\varepsilon n) + O(\varepsilon)$$

where

$$\frac{dU_0}{ds} = U_0 B$$

and

$$B = \frac{1}{N} \sum_{j=0}^{N-1} B_j$$

Therefore, the average influence of the sequence $\{B_n\}$ is dominant to leading order. The higher order approximations of this chain using the multiscale method bring the higher moments of the probability distribution of the sequence $\{B_n\}$ into the approximation. Although these results are not surprising, they are useful since it is frequently possible to estimate the small parameter ε from experiments [see, e.g., 7.8].

7.4   Almost Harmonic Systems

Problems of the form

$$\frac{dz}{dt} = Az + \varepsilon f(t, z, \varepsilon)$$

where A is an oscillatory matrix, ε is a small parameter, and f is a smooth function of its arguments are referred to as being *almost harmonic systems*. (Recall that A is an *oscillatory matrix* if it is a diagonalizable real matrix and all of its eigenvalues are purely imaginary, say $\pm i\omega_1, \cdots, \pm i\omega_N$)

We suppose that A is real, and so its dimension is even: say $z, f \in E^{2N}$ and $A \in E^{2N \times 2N}$. If f is a smooth function of its variables and is almost periodic in t, then this system can be converted into one to which the averaging theory applies directly by using the change of variables

$$z = \exp(At)y$$

With this,

$$\frac{dy}{dt} = \varepsilon e^{-At} f(t, e^{At}y, \varepsilon)$$

The new right-hand side has a generalized Fourier expansion in t, and the averaging theory can be tried directly as described earlier.

From another point of view, we can introduce a change of variables that takes the matrix A into its real Jordan canonical form. Since A is an oscillatory matrix, there is a matrix, say P, such that

$$P^{-1}AP = \text{diag}\left[\begin{pmatrix} 0 & \omega_1 \\ -\omega_1 & 0 \end{pmatrix}, \cdots, \begin{pmatrix} 0 & \omega_N \\ -\omega_N & 0 \end{pmatrix}\right]$$

When $\varepsilon = 0$, the problem reduces to one that is equivalent to a system of harmonic oscillators, which is why we refer to such systems for z as being almost harmonic.

We can rewrite the system for z as

$$\frac{d^2 x_j}{dt^2} + \omega_j^2 x_j = \varepsilon F_j\left(t, x, \frac{dx}{dt}, \varepsilon\right)$$

for $j = 1, \cdots, N$. For example, x_j can be thought of as the location of the jth particle in an N-particle array each with momentum dx_j/dt. Then this equation is based on Newton's law, and it describes the restoring force in the system ($\omega_j^2 x_j$) and the coupling between particles εF_j.

Almost harmonic systems of this form are studied in Refs. 7.9–7.13. We derive a variety of interesting results for such systems.

7.4.1   *Phase-Amplitude Coordinates*

One advantage of studying harmonic oscillators is that they can be easily transformed into phase-amplitude coordinates (see Section 2.3.5). Let us define variables r_j and θ_j by

$$r_j e^{i\theta_j} = \frac{dx_j}{dt} + i\omega_j x_j$$

Therefore,

$$\frac{dr_j}{dt} + i\frac{d\theta_j}{dt}r_j = \varepsilon e^{-i\theta_j}F_j + i\omega_j r_j$$

and so the new variables $r_j, \theta_j \in E^N$ satisfy the equations

$$\frac{dr_j}{dt} = \varepsilon f_j(t, r, \theta, \varepsilon)$$

$$\frac{d\theta_j}{dt} = \omega_j + \varepsilon g_j(t, r, \theta, \varepsilon)$$

where

$$f_j = \mathrm{Re}(\exp(-i\theta_j)F_j)$$

$$g_j = \mathrm{Im}(\exp(-i\theta_j)F_j)/r_j$$

The vector of amplitudes r and the vector of phases θ are the unknowns in this problem. This change of variables converts the original system into phase-amplitude coordinates.

The averaging procedures outlined in Section 7.1 can be applied to this system, which includes a number of interesting examples. Small divisor problems are common in these systems, but they can be eliminated in many useful cases. For example if the averaged system has an asymptotically stable rest point, then mean stability overcomes small divisors. Also, if the free frequencies of the system (ω_j) satisfy certain nonresonance conditions, then small divisors can be shown to be manageable.

7.4.2 Free Oscillations

When there is no external forcing in the system, the equations are

$$\frac{dr_j}{dt} = \varepsilon f_j(r, \theta, \varepsilon)$$

$$\frac{d\theta_j}{dt} = \omega_j + \varepsilon g_j(r, \theta, \varepsilon)$$

When $N = 1$ and there is no external forcing, we can use θ as the new time variable in the problem:

$$\frac{dr}{d\theta} = \varepsilon \frac{f(r, \theta, \varepsilon)}{\omega + \varepsilon g(r, \theta, \varepsilon)}$$

and treat this using our methods for forced systems. Once this is done, the additional step of relating θ to t must be taken. If $r = r(\theta, \varepsilon)$ is known, then

$$\frac{d\theta}{dt} = \omega + \varepsilon g(r(\theta, \varepsilon), \theta, \varepsilon)$$

The solution is $\theta = \theta(t, \varepsilon)$. For example, if the period of an oscillation that has period 2π in θ is to be found, we must solve the equation $2\pi = \theta(T, \varepsilon)$ for the period T. As we have seen before, the period usually changes as ε does.

When $N > 1$, since all of the phase variables are timelike ($\theta_j \sim \omega_j t + \cdots$), it is not obvious how perturbation methods can be useful.

One approach is to change the basis in E^N using $\omega = \mathrm{col}(\omega_1, \cdots, \omega_N)$ and its orthogonal complement, so $\omega, W_2, \cdots, W_N$, is an orthogonal basis. Then

$$\frac{d(\omega \cdot \theta)}{dt} = \omega \cdot \omega + \varepsilon \omega \cdot g(r, \theta, \varepsilon)$$

$$\frac{d(W_j \cdot \theta)}{dt} = \varepsilon W_j \cdot g(r, \theta, \varepsilon)$$

for $j = 2, \cdots, N$. Thus, we see that $\omega \cdot \theta$ is a timelike variable and all the other variables in the problem move slowly with respect to it. This approach is studied in the rotation vector method in Section 7.5.1, and it is successful if the conditions of the mean-stable averaging theorem are met by the system

$$\frac{dr_j}{d\tau} = \varepsilon \frac{f_j}{\omega \cdot \omega + \varepsilon \omega \cdot g}$$

$$\frac{d\psi_j}{d\tau} = \varepsilon \frac{W_j \cdot g_j}{\omega \cdot \omega + \varepsilon \omega \cdot g}$$

where $\psi_j = W_j \cdot \theta$ and $\tau = \omega \cdot \theta$.

On the other hand, if there is no strong stability and the frequencies ω_1, $\cdots$, ω_N are incommensurable, little is known (see Section 7.5.6.) The problem in this case can be partly solved in some cases if r can be shown to approach an invariant manifold, say $r \to R(\psi, \varepsilon)$ as $t \to \infty$, in which case. The problem reduces to one for ψ alone on this manifold, and so to a flow on a torus. The invariant torus can be stable, but behavior on it might be ergodic. This is an extension of the ideas of orbital stability (see Section 7.4.3).

If there is an oscillation for this system, then its period probably will change with ε, and so we must determine its period, or equivalently its frequency, as we construct a solution. Similar to our discussion of orbital stability of free oscillations, we attempt here to construct an orbit and its period, and we disregard the phase lag of solutions. Van der Pol's equation again gives a good illustration.

Let us consider the equation

$$\frac{d^2 x}{dt^2} + x = \varepsilon(1 - x^2)\frac{dx}{dt}$$

from the points of view of two time scales and of almost harmonic systems.

Two-scale approximation

We look for the solution of van der Pol's equation in the form

$$x = x_0(\tau, s) + x_1(\tau, s)\varepsilon + x_2(\tau, s)\varepsilon^2 + \cdots$$

whre s ($= \varepsilon t$) and $\tau = \omega(\varepsilon)t$ are taken to be independent variables. As before, we require that the coefficients in this expansion be bounded as functions of t, and to determine $\omega(\varepsilon)$ we require that they be 2π-periodic as functions of τ as well. Now, x satisfies the equation

$$\omega^2 \frac{\partial^2 x}{\partial \tau^2} + 2\varepsilon\omega \frac{\partial^2 x}{\partial \tau \partial s} + \varepsilon^2 \frac{\partial^2 x}{\partial s^2} + \varepsilon\omega(x^2 - 1)\left(\frac{\partial x}{\partial \tau} + \varepsilon \frac{\partial x}{\partial s}\right) + x = 0$$

Setting $\varepsilon = 0$ in this equation gives

$$\omega_0^2 \frac{\partial^2 x_0}{\partial \tau^2} + x_0 = 0$$

The general solution of the first equation is

$$x_0(\tau, s) = A(s)\cos \tau + B(s)\sin \tau$$

where A and B are arbitrary functions of s, and we see from the periodicity condition that $\omega_0 = 1$. It will turn out that we must take $\omega_1 = 0$ to ensure that x_1 is periodic. We use this information at this point to simplify our calculations. Differentiating with respect to ε and setting $\varepsilon = 0$ gives

$$\frac{\partial^2 x_1}{\partial \tau^2} + x_1 = -2 \frac{\partial^2 x_0}{\partial \tau \partial s} + (1 - x_0^2)\frac{\partial x_0}{\partial \tau}$$

The equation for x_1 becomes

$$\frac{\partial^2 x_1}{\partial \tau^2} + x_1 = -2(-A'\sin \tau + B'\cos \tau)$$

$$+ (1 - A^2 \cos^2 \tau - AB\sin 2\tau - B^2 \sin^2 \tau)(-A\sin \tau + B\cos \tau)$$

Using the identities

$$\cos^3 t = (3\cos t)/4 + \text{h.f.t.}$$

$$\sin^3 t = (3\sin t)/4 + \text{h.f.t.}$$

$$\cos^2 t \sin t = (\sin t)/4 + \text{h.f.t.}$$

$$\sin^2 t \cos t = (\cos t)/4 + \text{h.f.t.}$$

we can easily expand the right-hand side of this equation in its Fourier series.

The functions A and B must be chosen to suppress the resonant terms in the right-hand side (see Chapter 6). To do this, the coefficients of $\cos \tau$ and $\sin \tau$ are set equal to zero, and the result is

$$2\frac{dA}{ds} - A\left(1 - \frac{A^2}{4} - \frac{B^2}{4}\right) = 0$$

$$2\frac{dB}{ds} - B\left(1 - \frac{A^2}{4} - \frac{B^2}{4}\right) = 0$$

This system can be solved using polar coordinates: Setting $A = R\cos \psi$ and

$B = R \sin \psi$, we have that

$$2R \frac{dR}{ds} = 2A \frac{dA}{ds} + 2B \frac{dB}{ds} = R^2 \left(1 - \frac{R^2}{4} \right)$$

and

$$R^2 \frac{d\psi}{ds} = A \frac{dB}{ds} - B \frac{dA}{ds} = 0$$

Therefore,

$$\frac{dR^2}{ds} = R^2 \left(1 - \frac{R^2}{4} \right)$$

We see that $R(\infty) = 2$, so the result of these computations is that as $t \to \infty$,

$$x \to 2 \cos(\omega t + \phi) + O(\varepsilon)$$

where $\omega = 1 + O(\varepsilon^2)$ and ϕ is determined by initial conditions. This approach to the averaging method is used extensively in Refs. 7.14 and 7.15, which contain many interesting examples showing how the method is used in a variety of modern applications.

Systems of van der Pol's equations have the form

$$\frac{d^2 x_j}{dt^2} + \mu_j (x_j^2 - 1) \frac{dx_j}{dt} + \omega_j^2 x_j = \varepsilon f_j \left(t, x, \frac{dx}{dt} \right)$$

where ε and μ_j are parameters, and x and dx/dt denote vectors with components $x_1, \cdots, x_N$, and $dx_1/dt, \cdots, dx_N/dt$, respectively.

When $\varepsilon = 0$, this system reduces to a system of van der Pol's equations with parameters μ_j and frequencies ω_j. We have just shown that if each $0 < \mu_j \ll 1$, then each oscillator approaches a periodic solution having amplitude $R = 2$. Since each equation in this system has a stable periodic solution, the entire network has a stable invariant torus that is the product of these orbits. In Ref. 7.16, it is shown that there is a nearby invariant torus for the full system when $0 < \varepsilon \ll 1$.

7.4.3 *Conservative Systems*

Recall that Hamiltonian systems can be (in theory) converted to phase-amplitude coordinates where an amplitude variable is the system's energy and the phase variables are defined in a complicated way (See Section 2.3.5).

Let us consider a system that has been converted to phase-amplitude coordinates, say in the form

$$\frac{dr}{dt} = f(r, \psi, \varepsilon)$$

$$\frac{d\psi}{dt} = g(r, \psi, \varepsilon)$$

where r is a vector of N amplitude variables, ψ is a vector of N angular phase variables, and f and g are smooth functions that are 2π-periodic in each of the components of ψ.

We suppose that $r = r*$ corresponds to the invariant torus for $\varepsilon = 0$, so

$$f(r*, \psi, 0) = 0$$

for all ψ. Then the set $(r*, \psi)$ defines an N torus in E^{2N} as the components of ψ range over $[0, 2\pi]$. Moreover, the equation

$$\frac{d\psi}{dt} = g(r*, \psi, 0)$$

defines a flow of solutions on the surface of the torus. Linearizing the r equation about $r = r*$ and replacing $r - r*$ by εr brings the system into the form

$$\frac{dr}{dt} = Sr + O(\varepsilon)$$

$$\frac{d\psi}{dt} = g(r*, \psi, 0) + O(\varepsilon)$$

The special case

$$g(r*, \psi, 0) = \omega(r*)$$

where $\omega(r*)$ is a function of $r*$ only, is usually the one considered, since little is known about the general case. For most of our work, we take ω to be constant. Note that this system is of the same form as that derived for the weakly coupled harmonic system except that r can now respond on the fast time scale t.

We would like to derive an approximation to solutions of this system that is valid uniformly for $0 \le t < \infty$. However, we cannot use the mean-stable averaging theorem since it would require the ψ variable to equilibrate. This is usually not the case for flows on a torus, especially if the vector ω has incommensurable components. The following example illustrates this problem.

Recall our notation for quasiperiodic functions introduced in Chapter 1. Let $G(x_1, \cdots, x_N)$ be a smooth function that is 2π-periodic in each of the variables $x_1, \cdots, x_N$. Then if the frequencies $\omega_1, \cdots, \omega_N$ are rationally independent ($\mathbf{n} \cdot \omega \ne 0$ for all integer vectors $\mathbf{n} \ne 0$), the function $G(\omega_1 t, \cdots, \omega_N t)$ is a *quasiperiodic* function as described in Section 1.5.1. This is a special kind of almost-periodic function that is generated (or defined) by a periodic function of a finite number of variables. We say that its frequencies are spanned by ω. In particular, it has an easily derived Fourier series, and all of its frequencies are linear integer combinations of the finite set $(\omega_1, \cdots, \omega_N)$. The Fourier series of G has the form

$$G(\omega_1 t, \cdots, \omega_N t) = \sum_{|j| = -\infty}^{\infty} C_j e^{ij \cdot \omega t}$$

where $\mathbf{j}$ is a multiindex of integers, $\mathbf{j} = (j_1, \cdots, j_N)$, is the vector $\boldsymbol{\omega} = (\omega_1, \cdots, \omega_N)$ and $|\mathbf{j}| = j_1 + \cdots + j_N$.

Small divisor problems arise even in simple systems of weakly coupled harmonic oscillators. For example, consider the system

$$\frac{d^2x}{dt^2} + x = \varepsilon \sin(x - y)$$

$$\frac{d^2y}{dt^2} + 2y = \varepsilon \sin(y - x)$$

We will attempt to use the iteration method to construct oscillatory solutions of this system (see Section 6.2.2).

Setting $\varepsilon = 0$ gives

$$\frac{d^2x_0}{dt^2} + x_0 = 0$$

$$\frac{d^2y_0}{dt^2} + 2y_0 = 0$$

The solutions are

$$x_0(t) = A \cos t + B \sin t$$

$$y_0(t) = a \cos \sqrt{2}t + b \sin \sqrt{2}t$$

for some constants A, a, B, and b.

Putting these solutions into the right-hand side of the equation gives

$$\frac{d^2x_1}{dt^2} + x_1 = \sin(x_0 - y_0)$$

$$\frac{d^2y_1}{dt^2} + 2y_1 = \sin(y_0 - x_0)$$

Now, the right-hand side is a quasiperiodic function of t and it has a generalized Fourier series

$$\sin(x_0(t) - y_0(t)) = \sum_{m,n=-\infty}^{\infty} C_{m,n} \exp(imt + i\sqrt{2}nt)$$

where the coefficients $C_{m,n}$ are not zero for infinitely many values of m and n.

The solution for x_1 is given by the variation of constants formula

$$x_1(t) = \varepsilon e^{it} \int_0^t \sum_{m,n} C_{m,n} \exp(imt' + in\sqrt{2}t' - it') \, dt'$$

$$+ \varepsilon e^{-it} \int_0^t \sum_{m,n} C_{m,n} \exp(imt' + in\sqrt{2}t' + it') \, dt' + \alpha e^{it} + \beta e^{-it}$$

Proceeding as though all sums converge, we have that

$$x_1(t) = \varepsilon \sum_{m,n} C_{m,n} \frac{e^{imt+in\sqrt{2}t} - e^{it}}{im + in\sqrt{2} - i} + \sum_{m,n} C_{m,n} \frac{e^{imt+in\sqrt{2}t} - e^{-it}}{im + in\sqrt{2} + i}$$

The following lemma [7.17] highlights a problem with this expansion.

Lemma. The set $\{1 + m + n\sqrt{2} : m, n \text{ integers}\}$ is dense in E^1.

Because of this fact, we see that the terms defining x_1 have arbitrarily small divisors, and we cannot expect these series to converge. It therefore appears that the iteration method fails to produce a quasiperiodic solution, and a subtler approach is needed to resolve this problem of small divisors.

Since $\sin(x - y)$ is a real analytic function of x and y, we do know that the coefficients $C_{m,n}$ decay geometrically to zero: Say, for some number ξ,

$$|C_{m,n}| \le |\xi|^{m+n}, \qquad |\xi| < 1$$

as $m + n \to \infty$. Therefore, if the divisors are not too small, then these series can be shown to converge, and the iteration process can be continued. It is possible to set conditions on the divisors that ensure that these series converge. The following definition is useful for dealing with small divisors.

Condition L. Let $(\omega_1, \cdots, \omega_N)$ be a set of frequencies. We say that these frequencies satisfy *condition L* if there are constants G and T with $G > 0$ and $T > N - 1$ such that for all multiindices $\mathbf{j} = (j_1, \cdots, j_N)$ with $\sum |j_k| \ne 0$ the estimate

$$|\boldsymbol{\omega} \cdot \mathbf{j}| > G |\mathbf{j}|^{-T}$$

holds.

This condition is named for Liouville who derived a similar condition. This condition is very much like a nonresonance condition (see Section 7.5.6). In fact, condition L holds for the example of this section [7.18], and so we see that the terms in the series defining $x_1(t)$ are dominated by those of a convergent geometric series. Therefore, x_1 is determined in the iteration method, and the process can continue.

When $\boldsymbol{\omega}$ is proportional to a vector of integers, the problem is said to be resonant and condition L is not satisfied. This is a case to which the Rotation Vector Method can be applied as shown in Section 7.5.1.

Bogoliuboff [7.12] introduced a method of near-identity transformations to analyze this system with condition L. The result is described by the following steps.

We introduce new variables by the transformation

$$\psi = u + \varepsilon U(u, v, \varepsilon)$$

$$r = \varepsilon V(u, v, \varepsilon)$$

where U and V are 2π-periodic in the components of u. This transformation is to be found, so that it takes the system into the simpler system

$$du/dt = \omega$$

$$dv/dt = Sv$$

Thus, if the matrix S is asymptotically stable, then the v values equilibrate to $v = 0$, and a solution in original variables will be

$$\psi = \omega t + c + \varepsilon U(\omega t + c, 0, \varepsilon)$$

$$r = \varepsilon V(\omega t + c, 0, \varepsilon)$$

Therefore, we see that such a solution is a quasiperiodic function of t.

The idea behind this transformation is to construct an invariant surface for $\varepsilon \neq 0$ that is near the invariant torus of the unperturbed system. Because of this, the method is sometimes referred to as being the *method of invariant manifolds*.

Since the problem is time invariant, we expect that a modification of frequencies will be required to carry out this plan. Therefore, we introduce a frequency modification λ, and we consider the modified problem

$$\frac{dr}{dt} = Sr + \varepsilon f(r, \psi, 0)$$

$$\frac{d\psi}{dt} = \omega + \lambda + \varepsilon g(r, \psi, 0)$$

Bogoliuboff derived the following theorem for this system.

Bogoliuboff's Averaging Theorem. *Suppose that the frequency vector ω satisfies condition L and that S is an asymptotically stable matrix. Finally, suppose that f and g are analytic functions of their arguments. Then there exists a smooth function $\lambda = \lambda(\varepsilon)$ such that the system above has a quasi-periodic solution of the form*

$$\psi = \omega t + c + \varepsilon U(\omega t + c, 0, \varepsilon)$$

$$r = \varepsilon V(\omega t + c, 0, \varepsilon)$$

where the phase shift c is an arbitrary vector of constants.

This theorem will be used decisively in Section 7.5.2 to construct intervals of phase locking. Unfortunately, it is not always easy to verify the conditions of this theorem.

7.5 Angular Phase Equations

In this section, we consider problems that are described in terms of phase variables alone. For example, these may arise as descriptions of dissipative or conservative flows on invariant tori or as models of networks of integrated circuits that are described by phase variables, as we saw for the **VCON** circuit.

The rotation vector method is useful for studying surprising phenomena of dissipation that occur within conservative systems. This includes phase-locking behavior for various systems.

Next, we apply Bogoliuboff's near-identity transformation to study a problem to which Denjoy's theory applies. This proves certain facts about phase locking. Finally, we describe briefly the KAM theory of flows on tori.

7.5.1 *Rotation Vector Method*

An important system of equations is

$$\frac{dx}{dt} = \omega + \varepsilon f(x, y, \varepsilon)$$

$$\frac{dy}{dt} = \varepsilon g(x, y, \varepsilon)$$

(7.2)

where y and g are M vectors and x, ω, and f are N vectors, f and g are periodic in each of the components of x, and ε is a small parameter. The functions f and g are assumed to be smooth functions of their arguments.

We suppose that ω is proportional to a vector of nonzero integers, and without loss of generality, we take

$$\omega = \text{col}(p_1, \cdots, p_N)$$

where p_k is an integer for each $k = 1, \cdots, N$. This system does not satisfy condition L, and so Bogoliuboff's averaging result does not apply directly.

However, vectors of integers $W_2, \cdots, W_N$ can be found that are orthogonal to ω and pairwise orthogonal:

$$W_j \cdot W_k = 0 \qquad \text{for } j \neq k$$

$\omega \cdot W_j = 0$ for $j = 2, \cdots, N$ and $\det(\omega, W) = 1$. We introduce new variables to the problem through a change of basis in E^N. Let

$$x = v\omega + \sum_{j=2}^{N} u_j W_j$$

where the new coordinates are $v, u_2, \cdots, u_N$.

This change of variables is similar to those called *Delaunay* orbit elements in celestial mechanics [7.19]. The differential equations are easily transformed into these new coordinates by taking dot products of the system with $\omega, W_2, \cdots, W_N$, successively. The results are

$$\frac{dv}{dt} = \omega \cdot \omega + \omega \cdot f$$

$$\frac{du_j}{dt} = \varepsilon W_j \cdot f$$

$$\frac{dy}{dt} = \varepsilon g$$

for $j = 2, \cdots, N$. We write $\omega^2 = \omega \cdot \omega$. Using v as a timelike variable, we divide the last $M + N - 1$ equations by the first and get

$$\frac{du_j}{dv} = \varepsilon \frac{W_j \cdot f}{\omega^2 + \omega \cdot f}$$

$$\frac{dy}{dv} = \varepsilon \frac{g}{\omega^2 + \omega \cdot f}$$

This form of the system is one to which the Mean-Stable Averaging Theorem can be applied.

We rewrite it as

$$\frac{du}{dv} = \varepsilon F(v, u, y, \varepsilon)$$

$$\frac{dy}{dv} = \varepsilon G(v, u, y, \varepsilon)$$

(7.3)

where u denotes the $N - 1$ vector $(u_2, \cdots, u_N)$ and F denotes the vector whose jth component is

$$F_j(v, u, y, \varepsilon) = \frac{W_j \cdot f}{\omega^2 + \omega \cdot f}$$

Since the original function f is periodic in the components of x and the vectors $\omega, W_2, \cdots, W_N$, have integer coordinates, the function F is periodic in v and the components of u. Thus, F and G have period 2π in v and the components of u. Therefore, F has a convergent Fourier series:

$$F(v, u, y, \varepsilon) = \sum_{|n| = -\infty}^{\infty} C_n(y, \varepsilon) \exp \left[i \left(n_1 v + \sum_{j=2}^{N} n_j u_j \right) \right]$$

The average of this function over v is given by the Fourier series

$$\langle F \rangle (U, y) = \sum_{|n| = -\infty}^{\infty} C_n(y, 0) \exp \left(i \sum_{j=2}^{N} n_j U_j \right)$$

where the sum is taken over all multiindices n for which $n_1 = 0$.

If the original function f has the Fourier series

$$f(x, y, \varepsilon) = \sum_{|n| = -\infty}^{\infty} f_n(y, \varepsilon) e^{in \cdot x}$$

then the average of f over v is given by the Fourier series

$$\langle f \rangle = \sum_{\substack{|n| = -\infty \\ n \cdot \omega = 0}}^{\infty} f_n(y, 0) e^{in \cdot x}$$

where the sum is taken over all multiindices n for which $n \cdot \omega = 0$. For such indices, we have that

$$n \cdot x = \sum_{j=2}^{N} u_j n \cdot W_j$$

Therefore, the coefficients C_n in the expansion for F can be determined from f.

Since the system is periodic in v, a near identity transformation can be found such that system (7.3) can be rewritten as

$$\frac{du}{dv} = \varepsilon \langle F \rangle (u, y) + O(\varepsilon^2)$$

$$\frac{dy}{dv} = \varepsilon \langle G \rangle (u, y) + O(\varepsilon^2)$$

(7.4)

We suppose that the following condition is satisfied.

Hypothesis H5. The averaged equation

$$\frac{dU}{dv} = \varepsilon \langle F \rangle (U, Y)$$

$$\frac{dY}{dv} = \varepsilon \langle G \rangle (U, Y)$$

has an exponentially stable rest point, say (U^*, Y^*).

This condition ensures that the state (U^*, Y^*) is stable under persistent disturbances. If this condition is satisfied, and if $u(0)$ is near U^*, $y(0)$ is near Y^*, and if ε is sufficiently small, then we have that

$$u = U + O(\varepsilon) \qquad \text{and} \qquad y = Y + O(\varepsilon)$$

uniformly for $0 \leq v < \infty$ for this system and all nearby (Caratheodory) systems as described in Section 8.1.

This calculation has an important application. Suppose that condition H5 is satisfied. Then the u components of the system remain bounded, and we have that

$$\frac{x}{v} = \omega + \sum_{j=2}^{N} \frac{u_j}{v} W_j \to \omega$$

as $v \to \infty$ for this system and all nearby ones as above. This can be written more concisely as

$$x_1 : \cdots : x_N \to \omega_1 : \cdots : \omega_N$$

meaning that the relative ratios of the components of x approach the relative ratios of frequencies ω. Thus, the output frequency ratios are locked. The u variables are, in engineering terminology, phase lags. Since they converge to near u^*, the phase lags also lock.

We say that the system is in phase-lock with the frequencies ω, and ω is called the *rotation vector* in analogy with the two-dimensional case described by Denjoy. These results are summarized in the next theorem.

Phase-Locking Theorem. *Suppose that f and g are smooth functions that are periodic in the components of x and that condition H5 is satisfied. Then*

$$x_1 : \cdots : x_N \to \omega_1 : \cdots : \omega_N$$

as $t \to \infty$ for system (7.2) and all nearby (Caratheodory) systems.

Phase-locking occurs in many conservative systems. This is due to dissipation within Hamiltonian systems. For example, if the system

$$\frac{dx}{dt} = f(x)$$

phase locks, say according to the rotation vector method, then $H(x, p) = f(x) \cdot p$ defines a Hamiltonian for which the dynamics are described by the equation

$$\frac{dx}{dt} = \text{grad}_p \, H(x, p) = f(x)$$

$$\frac{dp}{dt} = -\text{grad}_x \, H(x, p) = -\text{grad}(f(x) \cdot p)$$

In this way any system of phase equations can be embedded in a Hamiltonian system, so there are many conservative systems that have exponential stable subsystems.

7.5.2   *Rotation Numbers*

Rotation numbers are very useful for describing phase locking within systems of oscillators. For example, they can describe stability within systems where none is expected, such as for conservative systems in the preceding section. On the other hand, rotation numbers do not do everything: For example, recall the parabolic interval mapping in Section 2.5 (namely, $x \to rx(1 - x)$). We can define a "rotation number" to be the number of times an orbit circles around the rest point $(r - 1)/r$ as n increases, divided by n. Equivalently, we could count the number of times iterates move left in n steps and define the rotation number to be this number divided by n.

With this definition, there is rotation number $1/2$ for almost all initial values x_0 and for all values of r in the interval, $3 < r < r^* \sim 3.57 \cdots$. This happens even though there is an infinite sequence of period doubling bifurcations over this interval.

Still, rotation numbers can give useful information, ranging from Denjoy's rigorous theory for a flow on a torus, to Cartwright and Littlewood's [see

7.20–7.22] use of rotation numbers to study van der Pol's equation, even when this number is poorly defined.

7.5.3   Euler's Forward Method for Numerical Simulation

We begin this section with a useful observation based on our work on averaging for difference equations. Namely, we show that for a large class of problems Euler's forward difference method for solving angular phase equations is stable. In fact, it yields a first-order numerical scheme that gives the correct value for rotation number of a system that is in phase lock. We apply this simulation technique to two examples.

The forward Euler algorithm for solving a system of differential equations of the form

$$\frac{dx}{dt} = \omega + f(x, t)$$

where $x, \omega, f \in E^N$, and f is a smooth function that satisfies the conditions of the Nonlinear Averaging Theorem (Section 7.1.1), is given by

$$x_{n+1} = x_n + h(\omega + f(x_n, nh))$$

where we have used a step size h, replaced t by nh and the solution $x(t)$ by the sequence $x_n \sim x(nh)$. Setting

$$y_n = x_n - \omega nh$$

gives the formula

$$y_{n+1} = y_n + hf(y_n + hn\omega, nh)$$

and $y_0 = x_0$.

The difference between $y(nh)$ and y_n measures the error in using the algorithm. We say that a scheme is of order p if this difference error is $O(h^{p+1})$. The scheme for y_n is a first-order algorithm since

$$x(h) - x_1 = \int_0^h (\omega + f(x(s), s)) \, ds - h(\omega + f(x_0, 0))$$

$$= \int_0^h f(x(s), s) \, ds - hf(x_0, 0)$$

$$= O(h^2)$$

so $p = 1$.

According to the averaging theorem for difference equations (Section 7.3), we have

$$y_n = U(hn) + O(h)$$

where the function $U(s)$ is determined by solving the problem

$$\frac{dU}{ds} = \lim_{M \to \infty} \frac{1}{M} \sum_{j=0}^{M-1} f(U + \omega s, s) = f(U + \omega s, s)$$

and $U(0) = x_0$. To study this equation, we consider the averaged equation

$$\frac{d\overline{U}}{ds} = \lim_{S \to \infty} \frac{1}{S} \int_0^S f(\overline{U} + \omega s, s)\, ds = \bar{f}(\overline{U})$$

If this system has an exponentially asymptotically stable rest point, then U and $\overline{U}$ should remain close and $\overline{U}$ remains bounded. Therefore, if this happens

$$\frac{x_n}{nh} = \frac{y_n + nh\omega}{nh} \to \omega$$

as $n \to \infty$. This result agrees with the phase-locking results of Section 7.5.1. In particular, the condition used here is the same one used in the rotation vector method to ensure that the system phase-locks onto the rotation vector ω. Thus, we see that stability of this (doubly) averaged system ensures that the forward Euler algorithm generates a sequence that converges to the same rotation number as for the original system of equations.

This fact greatly facilitates our simulations of phase equations, since the forward Euler scheme involves few computations. We use it throughout the following examples.

7.5.4 *Computer Simulation of Rotation Vectors*

Coupled oscillators tend to synchronize—even Hamiltonian systems of them. This suggests that in addition to orbital stability (i.e., stability in amplitudes) there is significant phase-locking, which is asymptotic stability in angular phases. In fact, phase stability is the rule and erratic behavior, like that described by the KAM theory, is the exception in many physical systems.

The rotation vector method in Section 7.5.1 gives a rigorous method for demonstrating phase-locking. To illustrate this phenomenon using computer experiments, we consider two interesting examples.

$N = 2$

This case falls within the scope of Denjoy's theory. The rotation number described in Figure 7.2 is calculated using the forward Euler algorithm for the system

$$\frac{dx}{dt} = \omega - \varepsilon(\sin(x - y) + 2\sin(2x - y))$$

$$\frac{dy}{dt} = 1.0 + \varepsilon(\sin(x - y) + \sin(2x - y))$$

where $\varepsilon = 0.1$ and $\rho \sim x(50\pi)/y(50\pi)$. This shows plateau regions where the

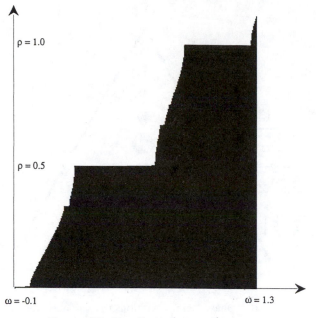

$\rho = 1.0$

$\rho = 0.5$

$\omega = -0.1$ $\omega = 1.3$

FIGURE 7.2. $N = 2$ rotation number vs. ω.

oscillators are in phase-lock. The KAM theory (Section 7.5.6) describes some of the system's behavior off of these plateaus.

$N = 3$. Three Oscillators.

The rotation vector in a system of three oscillators can be described using triangular coordinates. In the case of three phase equations, say

$$\frac{dx_j}{dt} = \omega_j + \varepsilon f_j(x_1, x_2, x_3)$$

for $j = 1, 2, 3$, if phase-locking occurs, the rotation vector will be the limit

$$\lim_{t \to \infty} x_1(t) : x_2(t) : x_3(t) = z_1 : z_2 : z_3$$

where the vector (z_1, z_2, z_3) is proportional to a vector of integers. We can represent this vector of integers using triangular coordinates by introducing numbers

$$\xi_j = \frac{z_j}{z_1 + z_2 + z_3}$$

for $j = 1, 2, 3$. Since these numbers add to one and they are not negative, we can depict the point (ξ_1, ξ_2, ξ_3) as being one in the intersection of the positive octant with the plane $\xi_1 + \xi_2 + \xi_3 = 1$. This is illustrated next for the model

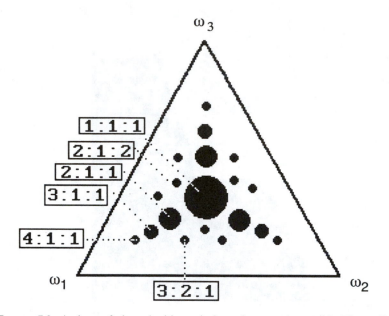

FIGURE 7.3. A chart of phase-locking relations that are observed in Figure 7.4.

$$\frac{dx_1}{dt} = \omega_1 + \varepsilon(\cos(x_1 - x_2) + \cos(2x_1 - x_3))_+$$

$$\frac{dx_2}{dt} = \omega_2 + \varepsilon(\cos(x_2 - x_3) + \cos(2x_2 - x_1))_+$$

$$\frac{dx_3}{dt} = \omega_3 + \varepsilon(\cos(x_3 - x_1) + \cos(2x_3 - x_2))_+$$

We plot the results of (ξ_1, ξ_2, ξ_3) on triangular (or areal) coordinates for each of many choices for $0 < \omega_1, \omega_2 < 2.0$. This is shown in Figures 7.3 and 7.4.

7.5.5   Near Identity Flows on $S^1 \times S^1$

Let us return to some work developed in our discussion of flows on a torus. Consider the equation

$$\frac{dx}{dt} = \omega(\varepsilon) + \varepsilon f(t, x, \varepsilon)$$

The change of variables $y = x - \omega(0)t$ takes this equation into one that has the form to which the averaging theory applies:

$$\frac{dy}{dt} = \varepsilon(\omega_1 + \omega_2\varepsilon + \cdots + f(t, y + \omega(0)t, \varepsilon))$$

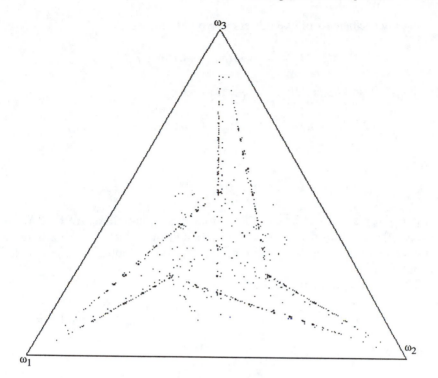

FIGURE 7.4. Simulation of three oscillators described in the text.

We can analyze this system using Bogoliuboff's near identity transformation: Let us try to find a change of variables

$$y = v(t, \xi, \varepsilon) = \xi + \varepsilon v_1(t, \xi) + \varepsilon^2 v_2(t, \xi) + \cdots$$

that takes the equation for y into one of the form

$$\frac{d\xi}{dt} = \varepsilon M_1(\xi) + \varepsilon^2 M_2(\xi) + \cdots$$

In this way, we have converted the original problem into one that does not involve t explicitly, much like the method of averaging converts the problem into a time invariant one. In fact, we should find that $M_1 = f_0$.

Substituting these formulas ($y = v(t, \xi, \varepsilon)$) into the problem for y gives

$$\frac{\partial v}{\partial t} + \frac{\partial v}{\partial \xi}\frac{d\xi}{dt} = \varepsilon[\omega_1 \varepsilon + \omega_2 \varepsilon^2 + \cdots + f(t, \omega_0 t + v, \varepsilon)]$$

Combining this formula with the ones for dx/dt and dv/dt gives

$$\frac{d\xi}{dt} + \varepsilon \frac{\partial v_1}{\partial t} + \varepsilon^2 \frac{\partial v_2}{\partial t} + \cdots + (\varepsilon M_1 + \varepsilon^2 M_2 + \cdots)\left(1 + \varepsilon \frac{\partial v_1}{\partial \xi} + \varepsilon^2 \frac{\partial v_2}{\partial \xi} + \cdots\right)$$

$$= \varepsilon f(t, \omega_0 t + v, \varepsilon) + \omega(\varepsilon) - \omega_0$$

Equating coefficients of like powers of ε on both sides gives

$$M_1 + \frac{\partial v_1}{\partial t} = f(t, \xi + \omega_0 t, 0) + \omega_1$$

so if $v_1(\xi, t)$ is to be periodic in t, we must have

$$M_1(\xi) = \omega_1 + \frac{1}{2\pi} \int_0^{2\pi} f(t, \xi + \omega_0 t, 0)\, dt \equiv \omega_1 + f_0(\xi)$$

and so

$$v_1(\xi, t) = V_1(\xi) + \int_0^t [f(t', \omega_0 t' + \xi, 0) - f_0(\xi)]\, dt'$$

where $V_1(\xi)$ is still not known. Two important points here are that M_1 is indeed the average of f (plus a constant), and the calculation here is quite similar of the two-time-scale expansion method introduced earlier in this chapter.

This process can continue. For example,

$$M_2 + M_1 \frac{\partial v_1}{\partial \xi} + \frac{\partial v_2}{\partial t} = \frac{\partial f}{\partial v}(t, \xi + \omega_0 t, 0) v_1(t, \xi) + f_1(t, \xi + \omega_0 t) + \omega_2$$

and so on (see [7.14, 7.15]).

7.5.6   KAM Theory

An important extension of Bogoliuboff's theorem was developed by Kolmogoroff, Arnol'd, and Moser [7.18], which is now known as the KAM theory. This work significantly relaxes the conditions needed by Bogoliuboff to establish the existence of quasiperiodic solutions to the system just above Bogoliuboff's averaging theorem in Section 7.4.3.

Suppose that the matrix S (Section 7.4.3) is diagonalizable and that its eigenvalues are denoted by σ_j. Condition L is now replaced by the following condition.

Condition KAM. Suppose that there are positive constants G and T, $T > N - 1$, such that

$$|i(\mathbf{j} \cdot \boldsymbol{\omega}) - \sigma_k| > G|\mathbf{j}|^{-T}$$

and

$$|i(\mathbf{j} \cdot \boldsymbol{\omega}) - \sigma_k + \sigma_{k'}| > G|\mathbf{j}|^{-T}$$

for all nonzero multiindices $\mathbf{j}$ as described in condition L and all indices k, $k' = 1, \cdots, m$.

The problem is modified by adding new variables μ, M, and λ as shown next

$$\frac{dr}{dt} = Sr + \mu(\varepsilon) + M(\varepsilon)r + \varepsilon f(r, \psi, \varepsilon)$$

$$\frac{d\psi}{dt} = \omega + \lambda(\varepsilon) + \varepsilon g(r, \psi, \varepsilon)$$

$$(7.5)$$

and we have the following theorem.

KAM Theorem. *Suppose that condition* KAM *is satisfied and that the functions f and g are analytic. Then there exist functions* $\lambda = \lambda(\varepsilon), \mu = \mu(\varepsilon),$ *and* $M = M(\varepsilon)$ *such that the system* (7.5) *has a quasiperiodic solution of the form*

$$\psi = \omega t + c + \varepsilon U(\omega t + c, \varepsilon)$$

$$r = \varepsilon V(\omega t + c, \varepsilon)$$

This result is useful in establishing the existence of quasiperiodic solutions to Hamiltonian systems. The proof of this result, a description of the earlier literature and several interesting examples are described in Ref. 7.13.

If λ, μ, and M in this result are zero, then we have found a quasiperiodic solution of the original system—in fact, an invariant torus of solutions. In any case, we see that if the nonresonance condition KAM is satisfied, then near the original system is one having a quasiperiodic solution.

7.6 Homogenization

We have considered averaging problems where two time scales appear, but there are many important problems where two space scales or combinations of time and space scales appear.

The idea here is to study periodic structures in space. Consider a material in which there is variation of physical parameters on both microscopic and macroscopic scales. For example, consider a line of bacterial cells that are fixed in an agar gel and placed at distances that are comparable to their size (microns). Suppose also that there is a gradient of nutrient that is established in the gel—its variation is on the macroscopic scale that is related to the size of the agar gel itself. Let a denote some physical characteristic, such as the local coefficient of diffusivity of nutrients, and let f be the local uptake rate of nutrients. Then, the data of the system can vary on the microscopic scale x/ε and the macroscopic scale x, where ε is the size of an interval containing each bacterial cell. We suppose that diffusivity is periodic with period ε, so we write $a(x/\varepsilon)$ where a has period 1.

A model for this is the one-dimensional diffusion equation

$$\frac{\partial}{\partial x}\left(a\left(\frac{x}{\varepsilon}\right)\frac{\partial u}{\partial x}\right) + f\left(x, \frac{x}{\varepsilon}\right)u = 0$$

$a(\xi)$ and $f(x, \xi)$ are periodic in ξ with period one. Setting

$$a\left(\frac{x}{\varepsilon}\right)\frac{\partial u}{\partial x} = v$$

then

$$\frac{\partial v}{\partial x} = -f\left(x, \frac{x}{\varepsilon}\right)u$$

This results in the system

$$\frac{\partial}{\partial x}\binom{u}{v} = \begin{pmatrix} 0 & b \\ -f & 0 \end{pmatrix}\binom{u}{v}$$

where $b = 1/a$. Using the two-scale method developed in Section 7.1, we replace $\partial/\partial x$ by $\varepsilon\partial/\partial x + \partial/\partial y$ where $y = \varepsilon x$. Then the equations become

$$\left(\varepsilon\frac{\partial}{\partial x} + \frac{\partial}{\partial y}\right)\binom{u}{v} = \varepsilon\begin{pmatrix} 0 & b \\ -f & 0 \end{pmatrix}\binom{u}{v}$$

Expanding u and v in terms ε, we have

$$u = u_0(x, y) + \varepsilon u_1(x, y) + \cdots$$

$$v = v_0(x, y) + \varepsilon v_1(x, y) + \cdots$$

Substituting these into the equation and equating like powers of ε gives

$$\frac{\partial}{\partial y}\binom{u_0}{v_0} = 0$$

so

$$\binom{u_0}{v_0} = \binom{U_0(x)}{V_0(x)}$$

and

$$\frac{\partial}{\partial y}\binom{u_1}{v_1} = \begin{pmatrix} 0 & b \\ -f_0 & 0 \end{pmatrix}\binom{u_0(x)}{v_0(x)} - \frac{\partial}{\partial x}\binom{u_0}{v_0}$$

As a result

$$\frac{\partial}{\partial x}\binom{U_0}{V_0} = \begin{bmatrix} 0 & \left\langle\frac{1}{a}\right\rangle \\ -f_0 & 0 \end{bmatrix}\binom{U_0(x)}{V_0(x)}$$

where

$$\bar{b} = \left\langle\frac{1}{a}\right\rangle = \lim_{T\to\infty}\frac{1}{T}\int_0^T \frac{1}{a(s)}\,ds = \int_0^1 \frac{ds}{a(s)}$$

and

$$f_0(x) = \int_0^1 f(x, s)\,ds$$

Therefore,

$$\left(\frac{\partial}{\partial x} \left\langle \frac{1}{a} \right\rangle^{-1} \frac{\partial}{\partial x} \right) U_0 + f_0(x) U_0 = 0$$

and we see that the "mean diffusivity" is the *harmonic mean* of the micro-scopic diffusivity of the system (see Ref. 7.23 for a detailed treatment of homogenization).

7.7   Computational Aspects of Averaging

Let us consider now a highly oscillatory system, say having the form

$$\frac{dx}{dt} = f\left(\frac{t}{\varepsilon}, x \right), \qquad x(0) = \xi$$

where f is a vector of functions that satisfy the conditions of the Nonlinear Averaging Theorem of Section 7.1. We know from that theorem that the solution of this problem can be written in the form

$$x = x_0(t) + \varepsilon x_1(t, t/\varepsilon) + O(\varepsilon^2)$$

where x_0 is determined as being the solution of the equation

$$\frac{dx_0}{d\tau} = \bar{f}(x_0) = \lim_{T \to \infty} \frac{1}{T} \int_0^T f(\tau, x_0) \, d\tau, \qquad x_0(0) = \xi$$

and x_1 is given by the formula

$$x_1(t, t/\varepsilon) = X_1(t) + \int_0^{t/\varepsilon} [f(\tau, x_0) - \bar{f}(x_0)] \, d\tau$$

Under the assumptions of the theorem, the integral in this formula is uniformly bounded, and X_1 is (as yet) an undetermined function.

We wish to find $x(h, \varepsilon)$ at some step $1 \gg h \gg \varepsilon$. To do this we use and compare two methods: computation of averages and extrapolation.

7.7.1   *Direct Calculation of Averages*

First, to take advantage of the averaging formula, we must evaluate x_0, which entails evaluation of the average of f. This can be done in (at least) three different ways.

1. Approximation of the Averaging Limit

Since the average exists, given a tolerance δ, we can find a constraint $T(\delta, x)$ such that

$$\left| \frac{1}{T_1} \int_0^{T_1} f \, d\tau - \frac{1}{T_2} \int_0^{T_2} f \, d\tau \right| < \delta$$

and

$$\left| \frac{1}{T_1} \int_0^{T_1} f \, d\tau - \bar{f}(x) \right| < \delta$$

for all T_1 and $T_2 \geq T$. If we can find such a constraint $T(\delta, x)$, then we can take as an approximation to $\langle f \rangle$ the integral

$$\bar{f}(x) = \frac{1}{T(\delta, x)} \int_0^{T(\delta, x)} f \, d\tau + O(\delta)$$

The tolerance δ is related to the accuracy of the numerical scheme that we hope to obtain from this work.

2. Second Difference Method

Let us define

$$V(T, x) = \int_0^T f(\tau, x) \, d\tau = \bar{f}(x) T + p(T, x)$$

which has the form of the average of f times T plus a function p that is almost periodic in T and has mean zero.

First, note that if we take the second difference of v with respect to T, we get

$$V(2T, x) - 2V(T, x) = p(2T, x) - 2p(T, x)$$

For any tolerance δ and any value of x, there is a δ translation number $\mathbf{T}(\delta)$ for p such that

$$p(2\mathbf{T}, x) - p(\mathbf{T}, x) - p(\mathbf{T}, x) + p(0, x) = O(\delta)$$

uniformly for x in some domain G. Therefore, if $\mathbf{T}$ is such a translation number, then

$$V(2\mathbf{T}, x) - 2V(\mathbf{T}, x) = O(\delta)$$

Therefore, by tabulating values of the second difference of the integral:

$$V(2T, x) - 2V(T, x)$$

we can find candidates for translation numbers. Unfortunately, the fact that this difference is small does not necessarily imply that $|p(T, x)|$ is small.

Of course in some cases this method works exactly, for example, if f is periodic in t. Experience with this method has been good. It fails when f has an unfortunate set of frequencies. However, the method proposed here proceeds by tabulating the values of

$$V(2T, x) - 2V(T, x)$$

and selecting the first value of T away from $T = 0$ for which the tolerance

$$|V(2T,x) - 2V(T,x)| < \delta$$

is met. For this value of T, we take

$$\bar{f}(x) = \frac{V(T,x)}{T}$$

Various tests can be performed to create a sense of security in this estimate, but it could be undecidable.

3. Ergodic Theorem

If f is a quasiperiodic function, say

$$f(\tau, x) = F(\omega_1 \tau, \cdots, \omega_L \tau, x)$$

where F is 2π-periodic in each of its first L variables, then we know from the weak ergodic theorem [7.24] that

$$\bar{f}(x) = \lim_{T \to \infty} \frac{1}{T} \int_0^T f(\tau, x)\, d\tau = \frac{1}{(2\pi)^L} \int_0^{2\pi} \cdots \int_0^{2\pi} F(\theta_1, \cdots, \theta_L, x)\, d\theta_1 \cdots d\theta_L$$

Therefore, we could evaluate the L-fold integral wherever needed. This however can turn out to be a major project in itself if $L \geq 4$.

7.7.2 Extrapolation

Once a satisfactory value of T is found, we return to the full problem, where we set

$$\varepsilon' = h/T$$

Then

$$2x\left(h, \frac{\varepsilon'}{2}\right) - x(h, \varepsilon') = x_0(h) + O(\delta h) + O\left(\left(\frac{h}{T}\right)^2\right)$$

On the other hand,

$$x(h, \varepsilon) = x_0(h) + O(\varepsilon)$$

Therefore,

$$x(h, \varepsilon) = 2x\left(h, \frac{\varepsilon'}{2}\right) - x(h, \varepsilon') + O(\varepsilon) + O(\delta h) + O\left(\left(\frac{h}{T}\right)^2\right)$$

This formula shows how the extrapolation formula depends on the choice of δ, T, and ε'. Usually $\varepsilon' \gg \varepsilon$, so solving the problems for $x(h, \varepsilon'/2)$ and $x(h, \varepsilon')$ are usually significantly faster than calculating $x(h, \varepsilon)$ directly. One of the interesting features of this approach, as with the QSSA method in Chapter 8, is that the result improves as the stiffness ($\sim 1/\varepsilon$) increases [see 7.25, 7.26].

7.8 *Exercises*

7.1. a. Show that if $f(t)$ is a periodic function or a trigonometric polynomial, then condition H2 in section 7.1 is satisfied. That is, show that

$$\int_0^\tau [f(t) - \langle f \rangle]\, dt$$

   is bounded for all $\tau \geq 0$.

   b*. If $\xi(t, \omega)$ is a stationary random process, then its average exists

$$\lim_{T \to \infty} (1/T) \int_0^T E\{|\xi(t, \omega)|\}\, dt \equiv \langle \xi \rangle$$

   and in addition

$$\lim_{T \to \infty} (1/T) \int_0^T \{\xi(t, \omega) - \langle \xi \rangle\}\, dt \equiv \eta(\omega)$$

   is a gaussian random variable [8.30]. Find such a process $\xi(t, \omega)$ that does not have a convergent Fourier series, even one having random coefficients. This example shows that generalized Fourier analysis is not sufficient to describe random processes.

7.2. Show that if condition H3 in section 7.1 is replaced by the system being stable under persistent disturbances, then the mean averaging theorem still holds.

7.3. Averaging linear systems: Let A be an oscillatory matrix.
   a. Evaluate $\langle B \rangle \equiv \langle e^{-At}B(t)e^{At} \rangle$ when B is a matrix of periodic function.
   b. Show that $[\langle B \rangle, A] = 0$ where $[A, C] \equiv AC - CA$ is the commutator or Poisson bracket of A and C.

7.4. Show that the two-time method can be used for linear systems by carrying out the first two steps in solving

$$dx/dt = \varepsilon C(t)x$$

   where C is a periodic function or a trigonometric polynomial of t.

7.5. a. Verify the induction argument used in the proof of the Difference Equation Averaging Theorem.
   b. Show that if all of the eigenvalues of $\langle B \rangle$ have negative real parts, then the conclusion of the difference equation averaging theorem in Section 7.3. holds uniformly for all n, $n = 1, 2, 3, \cdots$.

7.6. Verify the spectral decomposition of the cyclic matrix used in Section 7.3.2.

7.7. Apply the method of averaging to the system

$$\frac{dy}{dt} = \varepsilon \exp(-At)f(\exp(At)y, \varepsilon)$$

   when A is an oscillatory matrix and the components of $f(y, \varepsilon)$ are smooth functions of ε and analytic functions of the y variables. Here y, f are in E^N and A is in $E^{N \times N}$.

7.8. a. Rewrite the equation

$$u'' + \omega^2 u = \varepsilon F(u)$$

as an equivalent linear system in E^2 whose coefficient matrix (when $\varepsilon = 0$) has eigenvalues $\pm i\omega$. Convert this system to phase-amplitude coordinates using polar coordinates (r, θ). Write the result as a first-order equation using θ as the timelike variable.

b. Consider Duffing's equation

$$x'' + ax = \varepsilon b x^3.$$

Describe its solution by using the averaging procedure. Compare your answers with those obtained for Duffing's equation in Chapter 6 for similar parameter values.

c. Apply the near-identity transformation to the system in part a.

7.9. Let $z(t)$ be a complex valued function that satisfies the differential equation

$$\frac{dz}{dt} = i\omega z + (a - zz^*)z$$

where z^* denotes the complex conjugate of z. Describe the behavior of solutions for $z(t)$ when a and ω are real numbers.

7.10. The gene pool carried by a population of animals that have a genetic trait (one locus) of one of three possible types (two alleles) can be described by the difference equation

$$g_{n+1} = \frac{r_n g_n + s_n(1 - g_n)}{r_n g_n^2 + 2s_n g_n(1 - g_n) + t_n(1 - g_n)^2}$$

Suppose that there is slow selection; that is,

$$r_n = 1 + \varepsilon \rho_n, \qquad s_n = 1 + \varepsilon \sigma_n, \qquad \text{and} \qquad t_n = 1 + \varepsilon \tau_n$$

where ρ_n, σ_n, and τ_n are assumed to be oscillatory functions that have mean values

$$\langle \rho \rangle = \lim_{n \to \infty} \frac{1}{n} \sum_{u=0}^{n-1} \rho_u, \qquad \langle \sigma \rangle = \lim_{n \to \infty} \frac{1}{n} \sum_{u=0}^{n-1} \sigma_u \qquad \langle \tau \rangle = \lim_{n \to \infty} \frac{1}{n} \sum_{u=0}^{n-1} \tau_u$$

With these notations, apply the Nonlinear Difference Equation Averaging Theorem to analyze this system for small values of ε in the four cases $\sigma > \rho, \tau$; $\sigma < \rho, \tau$; $\rho < \sigma < \tau$; and $\tau < \sigma < \rho$.

7.11. Let $f(t, x, \varepsilon) = \cos(x - \omega t) + \cos t$ in Section 7.5.5. Carry out the near-identity transformation for the resulting equation. Show that if $\omega_1 = 0$ and $\omega_0 \neq \pm 1$, then $M_2 = 0$ and, if $\omega_0 = \pm 1$, $-M_2 = \pm \sin x/2 + \omega_2$.

7.12*. Consider the initial value problem

$$\frac{\partial u}{\partial t} + b\left(\frac{x}{\varepsilon}, \frac{t}{\varepsilon}, x, t, \varepsilon\right)\frac{\partial u}{\partial x} = 0, \qquad u(x, 0) = U(x)$$

for real variables x, t, and for $0 < \varepsilon \ll 1$. Let $\xi = x/\varepsilon$ and $\tau = t/\varepsilon$, and suppose (i) $U(x)$ and $b(\xi, \tau, x, t, \varepsilon)$ are smooth (at least C^1) functions of ξ, τ, x, t, and ε.

and (ii) $b(\xi + 1, \tau, \cdots) = b(\xi, \tau + 1, \cdots) = b(\xi, \tau, \cdots)$ for all values of ξ, τ, x, t, and ε. That is, b is doubly periodic in ξ and τ having period 1. Use the multiple-scale formalism to construct an approximation to the solution of this problem that is valid for small values of ε: Let

$$u(x, t) = u_0(\xi, \tau, x, t) + \varepsilon u_1(\xi, \tau, x, t) + \cdots$$

and determine equations for coefficients, which should be bounded for all values of ξ, τ.

Show that $u_0(x, t)$ solves

$$\frac{\partial u_0}{\partial t} + p(x, t)\frac{\partial u_0}{\partial x} = 0, \qquad u_0(x, 0) = U(x)$$

{Hint: Consider the ordinary differential equation

$$d\xi/d\tau = b_0(\xi, \tau, x, t)$$

in which x and t appear as parameters. It is known [3.16] that for any choice of initial data τ_0, ξ_0, the solution, say

$$\xi = X(\tau, \tau_0, \xi_0)$$

has the properties that
(i)   the limit

$$\lim_{\tau \to \infty} X(\tau, \tau_0, \xi_0)/\tau = p(x, t)$$

exists and is independent of (τ_0, ξ_0);
(ii) p is a continuous function of (x, t);
(iii) if p is a rational number, then any solution approaches a periodic solution which is a torus knot,
(iv) if p is irrational, then any solution is dense in the torus.}

7.13*. Consider the problem [7.29]

$$\frac{du}{dt} = \varepsilon(B(t)u + C(t)v)$$

$$\frac{dv}{dt} = Az + \varepsilon(D(t)u + E(t)v)$$

where each of $B(t)$, $C(t)$, $D(t)$, and $E(t)$ is a finite sum of periodic matrices and the matrix A is stable. Suppose that the average

$$\lim_{T \to \infty} \frac{1}{T}\int_0^T B(s)\, ds = 0$$

Show that any solution to this system can be constructed in the form

$$u = u_0(t, \varepsilon t, \varepsilon^2 t) + O(\varepsilon)$$

$$v = v_0(t, \varepsilon t) + O(\varepsilon)$$

by using the multitime method.

7.14*. Consider the integro-differential equation [7.27]

$$\frac{dx}{dt} = \varepsilon \left\{ \mu + \int_0^t K(t-\tau)(\sin(x(\tau)+\tau) + \sin x(\tau))\,d\tau \right\}$$

where the kernel K satisfies $|K(\tau)| \leq Ae^{-a\tau}$ for $0 \leq \tau < \infty$. Here $|\varepsilon| \ll 1, 0 < a$, and μ is a fixed constant. Use the multitime Ansatz

$$x = x_0(t,s) + \varepsilon x_1(t,s) + \cdots$$

where $s = \varepsilon t$, to show that $x_0(t,s) = X_0(s)$ which satisfies the equation

$$\frac{dX_0(s)}{ds} = \mu + K^* \sin X_0$$

where

$$K^* = \lim_{T \to \infty} \frac{1}{T} \int_0^T \int_0^t K(t-\tau)\,d\tau\,dt.$$

7.15*. Consider the system of nonlinear diffusion equations [7.28]

$$\frac{\partial u}{\partial t} = D_1 \frac{\partial^2 u}{\partial x^2} + \varepsilon\lambda(u^2 + v^2)u - \omega v$$

$$\frac{\partial v}{\partial t} = D_2 \frac{\partial^2 v}{\partial x^2} + \omega u + \varepsilon\lambda(u^2 + v^2)v$$

where D_1, D_2, and ω are positive constants and $\lambda(r) = 1 - r^2$. Identify slowly modulated periodic waves. {Hint: Introduce polar coordinates to u and v by $R^2 = u^2 + v^2$ and $\tan \Theta = u/v$, average the problem and look for (traveling wave) solutions of the result having the form $R(\varepsilon x - c\varepsilon^2 t)$, etc., where c is a constant to be determined.}

8
Quasistatic-State Methods

The appearance of several time scales in a problem can mean that various components of the system equilibrate at different rates. Rapidly responding components can try to reach some equilibrium, while the other components change hardly at all. It is not surprising that such perturbation problems can be studied using stability methods, because both deal with how solutions approach equilibria. These problems differ from those in Chapter 7 where oscillations occurred on a fast time scale relative to other changes. In this chapter, we study problems that try to equilibrate on a fast time scale while other components in the system change more slowly.

Gradient systems illustrate many of the methods developed in this chapter. Consider the system of equations

$$\frac{dy}{dt} = -\nabla G(y)$$

where y is in E^N and G maps E^N into E_+^1. The solutions of this equation either begin and remain at a critical point of G, say y^* where $\nabla G(y^*) = 0$, or they approach such an equilibrium. Since along such a solution

$$\frac{dG(y(t))}{dt} = -|\nabla G(y)|^2$$

the solution must approach a critical point of $G(y)$.

An analogous singular perturbation problem is

$$\varepsilon \frac{dy}{dt} = -\nabla G(y)$$

The parameter ε seems not to be essential since the change of variables $\tau = t/\varepsilon$ removes it from the equation:

$$\frac{dy}{d\tau} = -\nabla G(y) \tag{8.1}$$

Still, in terms of the original time scale t the variable τ approaches ∞ as

$\varepsilon \to 0^+$. Let $Y(\tau)$ be a solution of Eq. (8.1) for which $\nabla G(Y(0)) \neq 0$. Then $Y(\tau) \to y^*$ as $\tau \to \infty$, where y^* is a critical point of G: $\nabla G(y^*) = 0$. In terms of the original time scale, we have $y = Y(t/\varepsilon)$. Therefore, for each $t > 0$, $Y(t/\varepsilon) \to y^*$ as $\varepsilon \to 0^+$! This calculation illustrates a duality between stability and quasistatic state problems. Note that y^* need not be a minimum of G, but, in most applications, we restrict attention to initial data $y(0)$ that lie in the basin of attraction of a minimum.

This result is robust: Solutions of the perturbed system

$$\varepsilon \frac{dy}{dt} = -\nabla G(y) + \varepsilon g(t, y, \varepsilon)$$

where ε is a small positive number behave in a way similar to Eq. (8.1). Suppose that g satisfies Caratheodory's conditions (g is Lebesgue integrable with respect to t, continuously differentiable with respect to the components of y, continuous as a function of ε, and bounded). Just as in our discussion of stability under persistent disturbances, we see that the function $G(y)$ is almost a potential function for this system, but its derivative along solutions is

$$\frac{dG(y(t))}{dt} = -\frac{1}{\varepsilon}|\nabla G(y)|^2 + \nabla G \cdot g(t, y, \varepsilon)$$

This is negative for all values of y except those near $\nabla G(y) = 0$. Therefore, if $y(0)$ is not near such a point, then $y(t) \to y^* + O(\varepsilon)$ as $t \to \infty$ or $\varepsilon \to 0^+$.

It is interesting to note that a similar calculation can work even if g represents (large deviation) gaussian noise [8.30].

Next consider the system of equations

$$\frac{dx}{dt} = f(t, x, y, \varepsilon)$$

$$\varepsilon \frac{dy}{dt} = -\nabla G(y) + \varepsilon g(t, x, y, \varepsilon)$$

$$(8.2)$$

Let us suppose that y^* is a minimum of $G(y)$ and that $y(0)$ is near y^*, then as in the preceding paragraph, the values of the solution $y(t)$ will lie near y^* as well. Suppose also that f satisfies Caratheodory's conditions and that the solutions of

$$\frac{dx}{dt} = f(t, x, y^*, \varepsilon)$$

beginning in a bounded domain Ω remain there for $0 \leq t \leq T$. Then the manifold $\Omega x \{y = y^*\}$ is referred to as being a *quasistatic manifold* for the system. It is not invariant with respect to solutions, but solutions of the full system with $y(0)$ near y^* certainly stay near this manifold for small ε and $0 \leq t \leq T$.

Finally, consider the system

$$\frac{dx}{dt} = -\nabla F(x) + \varepsilon f(t, x, y, \varepsilon), \qquad x(0) = \xi$$

$$\varepsilon \frac{dy}{dt} = -\nabla G(y) + \varepsilon g(t, x, y, \varepsilon), \qquad y(0) = \eta$$

(8.3)

If $y(0)$ is near y^* (a minimum of G), if $x(0)$ is near x^* (a minimum of F), and if f and g satisfy Caratheodory's conditions for y near y^* and x near x^* for $0 \leq t < \infty$, then, for small positive values of ε, y quickly equilibrates to a neighborhood of y^* (a quasistatic manifold) and x eventually equilibrates to a neighborhood of x^*. In particular, the solution is near

$$x = x^* + \mathbf{x}(t) + O(\varepsilon)$$

$$y = y^* + Y(t/\varepsilon) + O(\varepsilon)$$

where

$$\frac{d\mathbf{x}}{dt} = -\nabla F(x^* + \mathbf{x}), \qquad \mathbf{x}(0) = \xi - x^*$$

and

$$\frac{dY}{d\tau} = -\nabla G(Y), \qquad Y(0) = \eta - y^*$$

This approximation is valid on the entire interval $0 \leq t < \infty$, and we see that it is a sum of terms: the equilibrium (x^*, y^*) of the reduced problem ($\varepsilon = 0$ in Eq. (8.3)), two transients one on the slow time scale t(for $\mathbf{x}$) and one on the fast time scale τ(for Y), and a small error. The expression $(x^* + \mathbf{x}, y^*)$ is the called the *quasistatic-state approximation* and Y is the initial transient.

It is clear that level sets of the potential functions F and G in system (8.3), say

$$\{(x, y)| F(x) = F(x^*) + a\varepsilon \qquad \text{and} \qquad G(y) = G(y^*) + b\varepsilon\}$$

where a and b are fixed constants, defines a tube in $[0, \infty) \times E^m \times E^n$ that is of order $o(1)$ about x^*, y^*. It is "sticky" in the sense that solutions of the system starting near it are attracted to it and once inside the boundary stay there.

The first example (8.1) describes a duality between perturbations and stability under persistent disturbances. The next example shows that behavior on various time scales can be peeled off. For example, the dynamics of the first of Eqs. (8.2) with $y = y^*$ and $\varepsilon = 0$ could be chaotic but still y is attracted to y^*. Finally, the third example shows how an approximation can be constructed for $0 \leq t < \infty$ when the slow and fast auxiliary problems following the system (8.3) are both stable. These examples are useful to keep in mind throughout this chapter, where various results are derived that extend and apply these observations.

The results just described for perturbed gradient systems can be extended to more general systems. In fact, a Liapunov function centered at some state is quite similar, at least locally, to a potential function. For example, consider the system

$$\varepsilon \, dy/dt = G(y)$$

and suppose that it has a static state, say $y = y^*$, that is asymptotically stable. Then from Chapter 3 we know that there is a Liapunov function $W(y)$ that is continuously differentiable at and near y^*, $W(y^*) = 0$, and there is a comparison function c such that

$$\frac{dW}{dt} = \frac{1}{\varepsilon} \nabla W \cdot G(y) \le -\frac{1}{\varepsilon} c(|y - y^*|)$$

for y near y^*. It follows that $W(Y(t/\varepsilon)) \to W(y^*)$ as $\varepsilon \to 0^+$ for every $t > 0$, and so $y \to y^*$ if $y(0)$ is near y^*.

We can apply W to study solutions of a nearby (Caratheodory) system

$$\frac{dy}{dt} = \frac{1}{\varepsilon} G + g(t, y, \varepsilon)$$

in which case $y(t, \varepsilon) \to y^* + o(1)$ as $\varepsilon \to 0^+$ for every $t > 0$ [8.1]. Liapunov functions are used later to extend the quasistatic-state approximation beyond gradient-like systems.

In Chapter 1, we saw that the spectrum of the matrix A determines the behavior of solutions of the linear system

$$dx/dt = Ax$$

In particular, as in Chapter 7, if the eigenvalues of A are real but widely separated, then the solution has relatively fast and slow modes. The ratios of the eigenvalues show what are the fast and slow time scales.

The term "singular perturbation problem" applies to any problem whose solution has a near-singular behavior, in the sense of complex function theory, relative to a perturbation parameter in the system. Usually the solutions have essential singularities at $\varepsilon = 0$. The problems studied in Chapter 7 are singular perturbation problems since their solutions involve terms of the form $\exp(it/\varepsilon)$. In this chapter, we study problems that have slowly varying parts and rapidly equilibrating parts—typically they include terms of the form $\exp(-t/\varepsilon)$. In contrast to the last chapter, we do not study highly oscillatory problems here, but only problems whose solutions are the sum of a (slow) quasistatic state (QSS) and a rapid initial transient to it.

Linear problems illustrate these results from the analytic point of view. For example, the equation

$$\frac{dx}{dt} = \frac{1}{\varepsilon} Ax + f(t)$$

has its solution given explicitly by the formula

$$x(t) = \exp\left(\frac{At}{\varepsilon}\right) x(0) + \int_0^t \exp\left(\frac{A(t-s)}{\varepsilon}\right) f(s)\, ds$$

If the eigenvalues of A are real and negative, and if ε is a small positive number, then the kernel $\exp(A(t-s)/\varepsilon)$ behaves like a delta function in the sense that

$$x(t) = e^{At/\varepsilon}(x(0) + \varepsilon A^{-1} f(t)) - \varepsilon A^{-1} f(t) + O(\varepsilon^2 A^{-2})$$

The first and last terms are small, so, as a useful approximation, we write

$$e^{At/\varepsilon} \sim -\varepsilon A^{-1} \delta(t)$$

for $t \geq 0$ and ε near zero. This calculation also shows that the solution $x(t)$ is approximated by the sum of a rapidly decaying exponential $(\exp(At/\varepsilon))$ and a slowly varying term $(-\varepsilon A^{-1} f(t))$. Integrals of this form are frequently encountered in proofs of quasistatic-state approximations.

A system of differential equations, say

$$dz/dt = F(t, z)$$

where z in E^{m+n}, might have two time scales, say t and t/ε. It is often the case that the parameter ε can be identified and that the system can be converted to the form

$$\frac{dx}{dt} = f(t, x, y, \varepsilon)$$

$$\varepsilon \frac{dy}{dt} = g(t, x, y, \varepsilon)$$

where $x, f \in E^m$, $y, g \in E^n$, and ε is a small positive parameter. On the other hand, it is often neither possible nor desirable to preprocess the system to get it into this form since the same system might have different time scales at various points in the state space (z). We describe an example of this at the end of the chapter. However, we mainly study the preprocessed system in this chapter, since we have to start somewhere.

It is important to consider only dimensionless parameters ε in these problems; in fact, trivial changes of units, say from nanometers to meters, in a model should not change a solution's behavior. Dimensionless parameters are also important in models since they make possible comparison of results obtained in various experimental settings [8.24].

We begin with a geometric description of fast and slow time scales. This is followed by an analysis of quasistatic states in a linear feedback system that shows how responses on the several time scales can be "peeled off" in succession, where behavior on the fastest scale is determined first, then on the next fastest, and so on. Fully nonlinear initial-value problems are studied next. We show that the fastest time response might equilibrate to a quasistatic manifold. For example, the manifold might contain quasistatic chaotic dynamics for which no perturbation theory gives adequate approximate solutions. Still, the solutions remain near the quasistatic manifold. Additional conditions on the manifold dynamics are shown to ensure convergence of the solution of the perturbed problem on the interval $0 \leq t \leq T(\leq \infty)$. It is shown under further conditions that the solution of the full problem can be approximated using the method of *matched asymptotic expansions*. The method of matched asymptotic expansions (MAE) [8.2–8.10] can be used to construct a *quasistatic-state approximation* (QSSA) and its initial transients for this problem. Extensions of this approach to a variety of other applications are discussed in later sections.

A theory of singular perturbations of nonlinear oscillations is described next. This begins with Friedrichs and Wasow's theory of quasistatic oscillations and ends with a brief description of nearly discontinuous oscillations. Boundary-value problems are described next, and the chapter ends with a description of nonlinear stability methods for static and stationary solutions, an analysis of H_2-O_2 combustion, and some computational schemes based on the QSSA.

8.1 Some Geometrical Aspects of Singular Perturbation Problems

It helps to have in mind a few simple pictures of singular perturbation problems. For example, consider a system of two scalar equations

$$\frac{dx}{dt} = f(x, y), \qquad x(0) = \xi,$$

$$\varepsilon \frac{dy}{dt} = g(x, y), \qquad y(0) = \eta$$

where f and g are smooth functions and ε is a small positive parameter.

Setting $\varepsilon = 0$ in this system results in an algebraic equation and a differential equation:

$$\frac{dx}{dt} = f(x, y), \qquad x(0) = \xi$$

$$0 = g(x, y)$$

This problem is called the *reduced problem*.

Solving the algebraic equation

$$g(x, y) = 0$$

may pose significant difficulties, but we assume that there is a unique smooth function $y = \phi(x)$ such that

$$g(x, \phi(x)) = 0$$

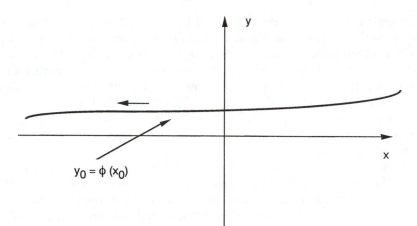

FIGURE 8.1. Solution of the reduced problem.

for all x. On this branch of solutions, the quasistatic problem becomes

$$\frac{dx}{dt} = f(x, \phi(x)), \qquad x(0) = \xi$$

We denote the solution of this problem by $x_0(t)$, and we write $y_0(t) = \phi(x_0(t))$. If $f < 0$, then x_0 moves to the left, as described in Figure 8.1.

Next, we see that the y component of the solution changes rapidly if g is not near zero since then

$$\left| \frac{dy}{dt} \right| = \frac{|g(x, y)|}{\varepsilon} \gg 1.$$

To investigate this further, we introduce the fast time variable $\tau = t/\varepsilon$, and the problem becomes

$$\frac{dx}{d\tau} = \varepsilon f(x, y) \qquad x(0) = \xi,$$

$$\frac{dy}{d\tau} = g(x, y) \qquad y(0) = \eta$$

Setting $\varepsilon = 0$ in this system gives

$$\frac{dx}{d\tau} = 0 \qquad x(0) = \xi,$$

$$\frac{dy}{d\tau} = g(x, y) \qquad y(0) = \eta$$

Equivalently, for $\varepsilon = 0$ and $x = \xi$ we have

$$\frac{dy}{d\tau} = g(\xi, y), \qquad y(0) = \eta$$

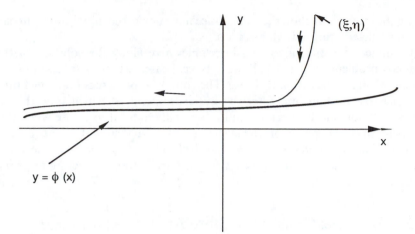

FIGURE 8.2. A nice singular perturbation problem. Arrows indicate the direction that solutions move in the (x, y) plane. Double arrows indicate fast motions.

This is called the *initial transient* problem. The value $y = \phi(\xi)$ is an equilibrium value for this equation, and, if it is stable, then we expect $y \to \phi(\xi)$ as $\tau \to \infty$. This happens if

$$g_y(\xi, \phi(\xi)) < 0$$

for all ξ, and then the solution of the full problem behaves as shown in Figure 8.2, which we refer to as being a nice singular perturbation problem.

In this chapter, we determine that some problems are nice singular perturbation problems, and we construct approximations to their solutions. The construction involves two steps: First, there is a special solution of the full problem that lies near $(x_0(t), y_0(t))$, called the *quasistatic state*. It can be found

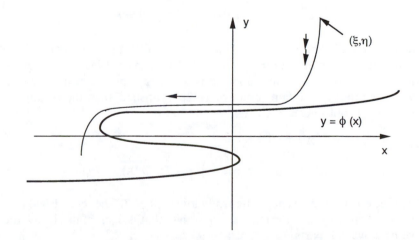

FIGURE 8.3. A not-nice singular perturbation problem.

using the implicit function theorem. Once it is known, the initial transient can be constructed using a Taylor's series.

Complications arise if g_y vanishes somewhere along the solution of the reduced problem. Figure 8.3 shows a typical case. A typical solution of the full problem is indicated in this case. The complication here is that a fast time response occurs sometime after the initial transient, at which time the reduced solution falls over a fold, and it is usually difficult to locate when this happens since it depends on the system's state. We deal with one aspect of these problems later in this chapter, but the fact that slow and fast time scales alternate in these problems poses a major barrier to constructing approximate solutions.

8.2   Example: Quasistatic State Analysis of a Linear Problem

The general idea is that the solutions of nice problems are the sum of a slowly varying part and a rapidly dying part, and each of these parts can be constructed separately. These can be combined to represent the full solution by matching free constants. The problems described in this section are ones for which the two parts of a solution can be constructed as power series in a single small parameter. Proofs are carried out in detail for a simple problem to indicate what steps are usually involved in validating QSSAs. Other problems are dealt with in Refs. 8.2, 8.3, 8.8, 8.9, and 8.11.

QSS methods can be illustrated using a linear problem, where the main ideas can be illustrated without technical complications. In particular, consider the linear problem

$$\frac{dx}{dt} = Ax + By + f(t) \qquad x(0) = \xi$$

$$\varepsilon\frac{dy}{dt} = Dy + g(t) \qquad\qquad y(0) = \eta$$

where $x, f, \xi \in E^m$, and $y, g, \eta \in E^n$. The matrices A, B, and D are constants of appropriate dimensions, $A \in E^{m \times m}$, $B \in E^{m \times n}$, and so on. Finally, we suppose that f and g have $N + 1$ continuous derivatives on some interval $0 \le t \le T$.

We show that if a certain condition (H2) is satisfied, then the solution of the original problem can be written as

$$x = x_0(t) + O(\varepsilon)$$

$$y = -D^{-1}g(t) + \exp\left(\frac{Dt}{\varepsilon}\right)(\eta + D^{-1}g(0)) + O(\varepsilon)$$

where the error estimates hold uniformly for $0 \le t \le T$. The result is valid for the case where $T = \infty$ if the matrix A is stable, that is, if all of its eigenvalues have negative real parts.

The quasistatic state is constructed using free constants ($\xi^*(\varepsilon)$) that are chosen at a later stage in the perturbation algorithm to ensure that matching conditions are met. The calculation of ξ^* shows that using the initial value ξ of the full problem does not necessarily give the correct quasistatic state beyond order ε.

8.2.1 Quasistatic Problem

The slowly varying part of the solution can be found by first solving the quasistatic problem and then later determining what initial conditions are appropriate for it. The quasistatic problem is posed in a curious way: Given any smooth function

$$\xi^*(\varepsilon) = \xi_0 + \varepsilon\xi_1 + \cdots + \varepsilon^N\xi_N + O(\varepsilon^N)$$

find functions x^* and y^*, such that

$$\frac{dx^*}{dt} = Ax^* + By^* + f(t), \qquad x(0) = \xi^*(\varepsilon)$$

$$\varepsilon\frac{dy^*}{dt} = Dy^* + g(t)$$

and x^*, y^* are smooth functions of ε for ε near zero. Note that (x^*, y^*) defines a solution of the system of differential equations, but no initial condition is specified for y^*; instead, we require that the solution (x^*, y^*) be expandable in a power series in ε.

The following calculation gives evidence that this problem is solvable for any choice of ξ^*. We set

$$x^*(t, \varepsilon) = x_0 + x_1\varepsilon + \cdots$$

$$y^*(t, \varepsilon) = y_0 + y_1\varepsilon + \cdots$$

and we study the problems that result for the coefficients in this expansion:

$$\frac{dx_0}{dt} = Ax_0 + By_0 + f(t), \qquad x_0(0) = \xi_0$$

$$0 = Dy_0 + g(t)$$

and

$$\frac{dx_1}{dt} = Ax_1 + By_1, \qquad x_1(0) = \xi_1$$

$$\frac{dy_0}{dt} = Dy_1$$

and so on. If D is invertible, each of these problems has a unique solution. Slightly more is needed to make the whole process valid.

Hypothesis H1. Suppose that the matrix D can be decomposed as

$$D = D_S + \sum_{j=1}^{M} i\beta_j P_j + D_U$$

where D_s and $-D_U$ are stable and $\beta_j \neq 0$ for all $j = 1, \cdots, M$. There are M purely imaginary eigenvalues. Moreover, we suppose that this decomposition is orthogonal. That is, $D_s D_U = D_s P_j = 0$, and so on.

The numbers $i\beta_j$ account for the purely imaginary eigenvalues of D, and a consequence of condition H1 is that the oscillatory part of D is diagonalizable. With this assumption

$$y_0 = -D^{-1}g$$

and x_0 is the unique solution of

$$\frac{dx_0}{dt} = Ax_0 + f - BD^{-1}g(t), \qquad x(0) = \xi_0$$

Once x_0 and y_0 are known, x_1 and y_1 are uniquely determined from the next two equations, and so on. This calculation is justified by the following theorem.

A Quasistatic-State Theorem. *Suppose that f, g, and ξ^* are smooth functions having continuous derivatives up to order $N + 1$ in their arguments for $0 \leq t \leq T$ ($T < \infty$) and for ε near zero. Moreover, suppose that D satisfies condition H1. Then there are functions (x^*, y^*) that solve the quasistatic problem and*

$$x^*(t, \varepsilon) = \sum_{j=0}^{N} x_j(t)\varepsilon^j + O(\varepsilon^{N+1})$$

$$y^*(t, \varepsilon) = \sum_{j=0}^{N} y_j(t)\varepsilon^j + O(\varepsilon^{N+1})$$

where the coefficients in this expansion are determined uniquely using the algorithm described above. The error estimate holds uniformly for $0 \leq t \leq T$.

Proof of the Quasistatic State Theorem. The remainders

$$R(t, \varepsilon) = x^*(t, \varepsilon) - \sum_{j=0}^{N} x_j(t)\varepsilon^j$$

and

$$Q(t, \varepsilon) = y^*(t, \varepsilon) - \sum_{j=0}^{N} y_j(t)\varepsilon^j$$

satisfy the equations

$$\frac{dR}{dt} = AR + BQ, \qquad R(0) = O(\varepsilon^{N+1})$$

$$\varepsilon \frac{dQ}{dt} = DQ + G$$

where

$$G = -D\frac{dy_N}{dt}\varepsilon^{N+1} = O(\varepsilon^{N+1})$$

The trick is in finding correct initial values for $Q(0)$. We write Q in terms of its projection onto the stable and oscillatory modes of D_1 (i.e., Q_1) and onto the unstable modes D_U (i.e., Q_U):

$$Q = Q_1 + Q_U$$

It is easy to verify that the formulas

$$Q_1(t) = \exp\left(\frac{D_1 t}{\varepsilon}\right)Q_1(0) + \frac{1}{\varepsilon}\int_0^t \exp\left(\frac{D_1(t-s)}{\varepsilon}\right)G_1(s)\,ds$$

$$Q_U(t) = -\frac{1}{\varepsilon}\int_t^T \exp\left(\frac{D_U(T-s)}{\varepsilon}\right)G_U(s)\,ds$$

define a solution of the remainder system. The function $Q_1(0)$ is determined explicitly by the algorithm, and $Q_U(0)$ is given explicitly by the last formula above. Finally, the functions G_1 and G_U are both of order $O(\varepsilon^{N+1})$, and integrating once by parts shows that

$$|Q_1(t)| \le K_0|Q_1(0)| + K_0\int_0^t |G_1(s)|\,ds$$

and

$$|Q_U(t)| \le K_1 \int_t^T |G_U(s)|\,ds$$

for some positive constants K_0 and K_1. Since $Q_1(0) = O(\varepsilon^{N+1})$, $Q(t) = O(\varepsilon^{N+1})$ uniformly for $0 \le t \le T$. Note that since $Q_1(0)$ is otherwise arbitrary, the quasistatic problem does not necessarily have a unique solution.

The solution for R is given by the formula

$$R(t) = e^{At}R(0) + \int_0^t e^{A(t-s)}BQ(s)\,ds = O(\varepsilon^{N+1})$$

In this example, the solution of the full problem for the remainders can be found explicitly. ∎

This result can be extended to the entire half-line $0 \le t < \infty$ if A is a stable matrix. In that case, the upper limit in the formula for Q_u is replaced by ∞, and the proof proceeds the same way.

8.2.2   Initial Transient Problem

We next derive an algorithm for determining ξ^* so that the differences

$$X(\tau, \varepsilon) = x - x^*, \qquad Y(\tau, \varepsilon) = y - y^*$$

are asymptotically equal to zero (up to order $N + 1$) for any $\tau > 0$ where $\tau = t/\varepsilon$. Moreover, we hope to find that $X(\tau, \varepsilon)$ and $Y(\tau, \varepsilon)$ are smooth functions of ε:

$$X(\tau, \varepsilon) = X_0(\tau) + \varepsilon X_1(\tau) + \cdots + X_N(\tau)\varepsilon^N + O(\varepsilon^{N+1})$$

$$Y(\tau, \varepsilon) = Y_0(\tau) + \varepsilon Y_1(\tau) + \cdots + Y_N(\tau)\varepsilon^N + O(\varepsilon^{N+1})$$

where the error estimates hold uniformly for $0 \le t \le T$. The functions X and Y are referred to as *initial transients*.

Since $\tau \to \infty$ as $\varepsilon \to 0^+$, X and Y will be asymptotic to zero as $\varepsilon \to 0^+$ for each $t > 0$ if the conditions

$$X_j(\tau) \to 0 \qquad \text{and} \qquad Y_j(\tau) \to 0$$

(exponentially) as $\tau \to \infty$ are met. These are sometimes called the *matching conditions* for the problem since they ensure that the QSS matches the initial transient.

The functions X and Y satisfy the system

$$\frac{dX}{d\tau} = \varepsilon(AX + BY), \qquad X(0) = \xi - \xi^*$$

$$\frac{dY}{d\tau} = DY, \qquad Y(0) = \eta - y^*(0, \varepsilon)$$

where $\xi^*(\varepsilon)$ is still unknown. Therefore,

$$\frac{dX_0}{d\tau} = 0, \qquad X_0(0) = \xi - \xi_0^*$$

$$\frac{dY_0}{d\tau} = DY_0, \qquad Y_0(0) = \eta + D^{-1}g(0)$$

It follows that $X_0(\infty) = 0$ only if $\xi_0^* = \xi$. This shows that there is a unique choice for ξ_0^*. In order that no restriction be placed on ξ and that $Y_0(\infty) = 0$, we require that

Hypothesis H2. D is a stable matrix (i.e., all of its eigenvalues have negative real parts, so $D = D_s$).

With this condition, $Y_0(\infty) = 0$ for any choice of (ξ, η). The functions X_j and Y_j solve the problems

$$\frac{dX_j}{d\tau} = AX_{j-1} + BY_{j-1}, \qquad X_j(0) = -\xi_j$$

$$\frac{dY_j}{d\tau} = DY_j, \qquad\qquad Y_j(0) = -y_j^*(0)$$

for $j = 1, 2, \cdots$. Therefore,

$$Y_j(\tau) = -\exp(D\tau)y_j^*(0)$$

and

$$X_j(\tau) = -\xi_j - BD^{-1}(e^{D\tau} - I)y_j^*(0) + A\int_0^\tau X_{j-1}(s)\,ds$$

To ensure that the matching condition on X_j is satisfied, we specify that

$$\xi_j = BD^{-1}y_j^*(0) + A\int_0^\infty X_{j-1}(s)\,ds$$

The integral in this formula exists since X_{j-1} approaches zero exponentially. This can be established by a straightforward induction argument, and so the expansion of $\xi^*(\varepsilon)$ is found.

Initial Transient Theorem. *Suppose that condition H2 is satisfied. Then there is a function $\xi^*(\varepsilon)$, uniquely determined to order $O(\varepsilon^{N+1})$ by ξ_j, $j = 0, \cdots, N$, such that the solution of*

$$\frac{dX}{d\tau} = \varepsilon(AX + BY), \qquad X(0) = \xi - \xi^*$$

$$\frac{dY}{d\tau} = DY, \qquad\qquad Y(0) = \eta - y^*(0, \varepsilon)$$

satisfies

$$X(t/\varepsilon, \varepsilon) = O(\varepsilon^{N+1})$$

and

$$Y(t/\varepsilon, \varepsilon) = O(\varepsilon^{N+1})$$

for each $0 < t \le T$ as $\varepsilon \to 0^+$. Moreover, these functions have uniquely determined Taylor expansions up to order $N + 1$, as derived in the last paragraph.

Proof of the Initial Transient Theorem. Let the functions X_j, Y_j be determined from the algorithm of the last paragraph, and let

$$R(t, \varepsilon) = X\left(\frac{t}{\varepsilon}, \varepsilon\right) - \sum_{j=0}^N X_j\left(\frac{t}{\varepsilon}\right)\varepsilon^j$$

$$Q(t, \varepsilon) = Y\left(\frac{t}{\varepsilon}, \varepsilon\right) - \sum_{j=0}^N Y_j\left(\frac{t}{\varepsilon}\right)\varepsilon^j$$

These functions satisfy the equations

$$\frac{dR}{dt} = AR + BQ \qquad R(0) = O(\varepsilon^{N+1})$$

$$\varepsilon \frac{dQ}{dt} = DQ \qquad Q(0) = O(\varepsilon^{N+1})$$

Integrating these equations, we have

$$Q(t) = \exp\left(\frac{Dt}{\varepsilon}\right) O(\varepsilon^{N+1})$$

and

$$R(t) = e^{At}O(\varepsilon^{N+1}) + \int_0^t e^{A(t-s')}e^{Ds'/\varepsilon}O(\varepsilon^{N+1})\,ds'$$

$$= O(\varepsilon^{N+1})$$

This completes the proof of the Initial Transient Theorem.   ∎

Again, note that the result is valid for $T = \infty$ if the matrix A is stable.

8.2.3 *Composite Solution*

The results of these calculations show that if condition H2 is satisfied, then the solution of the original problem can be written as

$$x(t, \varepsilon) = x^*(t, \varepsilon) + X(t/\varepsilon, \varepsilon)$$

$$y(t, \varepsilon) = y^*(t, \varepsilon) + Y(t/\varepsilon, \varepsilon)$$

where the quasistatic state (x^*, y^*) and the initial transient (X, Y) can be expanded in series in ε that are uniquely determined up to order ε^{N+1}. In particular,

$$x = x_0(t) + O(\varepsilon)$$

$$y = -D^{-1}g(t) + \exp\left(\frac{Dt}{\varepsilon}\right)(\eta + D^{-1}g(0)) + O(\varepsilon)$$

where the error estimates hold uniformly for $0 \le t \le T$. The result is valid for the case where $T = \infty$ if the matrix A is stable.

In summary, the quasistatic state is constructed using free constants $(\xi^*(\varepsilon))$ that are chosen at a later stage to ensure that the matching conditions are met. The calculation of ξ^* shows that using the initial value ξ of the full problem is not necessarily the correct choice for the quasistatic state beyond the first term.

8.2.4   *Volterra Integral Operators with Kernels Near δ*

The calculation in the proof of the Initial Transient Theorem highlights an important aspect of singular perturbations. Namely, the integrals involved have kernels that are quite like delta functions. To see this, consider the Volterra integral formula

$$g(t) = \frac{1}{\varepsilon} \int_0^t k\left(\frac{t-s}{\varepsilon}\right) f(s)\, ds$$

where the kernel is k and ε is a small positive parameter. We suppose that f and f' are smooth and bounded functions and that $k(t)$ and $K(t) = \int_t^\infty k(s)\, ds$ are both integrable functions. Then g can be approximated integrating by parts

$$g(t) = \frac{1}{\varepsilon} \int_0^t k\left(\frac{s}{\varepsilon}\right) f(t-s)\, ds = K(0)f(t) + \int_0^t K\left(\frac{s}{\varepsilon}\right) f'(t-s)\, ds$$

so

$$g(t) = K(0)f(t) + o(1)$$

Thus, we write $(1/\varepsilon)k((t-s)/\varepsilon) \sim K(0)\delta(t-s)$.

8.3   Quasistatic-State Approximation for Nonlinear Initial-Value Problems

Consider the initial-value problem

$$\frac{dx}{dt} = f(t, x, y, \varepsilon) \qquad x(0) = \xi(\varepsilon)$$

$$\varepsilon\frac{dy}{dt} = g(t, x, y, \varepsilon) \qquad y(0) = \eta(\varepsilon)$$

(8.4)

where $x, f \in E^m$, $y, g \in E^n$, and ε is a small positive parameter.

We consider a domain $\Omega = I \times B_R \times B_{R'} \times [0, \varepsilon_0]$ where $I = \{t: t_0 \leq t \leq T \leq \infty\}$, $B_R = \{x \text{ in } E^m: |x| \leq R\}$ and $B_{R'} = \{y \text{ in } E^n: |y| \leq R'\}$. T and ε_0 are some fixed constants. In what follows, the balls B_R and $B_{R'}$ can be replaced by any sets that are diffeomorphic to them. We suppose next that the following condition holds.

Hypothesis H3. f and g are $C^2(\Omega)$, and any solution of the system (8.4) beginning in $B_R \times B_{R'}$ remains there for $t_0 \leq t \leq T$.

Setting $\varepsilon = 0$ in this system gives

$$\frac{dx}{dt} = f(t, x, y, 0) \qquad x(0) = \xi(0)$$

$$0 = g(t, x, y, 0)$$

(8.5)

which we refer to as being the *reduced problem*. In the following subsection, we place successively more restrictive conditions on the system and obtain as a result more information about solutions.

8.3.1 *Quasistatic Manifolds*

We begin with the least restrictive result.

Hypothesis H4. Suppose there is a function $y = \Phi(t, x)$ such that $g(t, x, \Phi(t, x), 0) = 0$ for $t_0 \leq t \leq T$ and x in B_R. Moreover, this function is smooth, Φ in $C^2(I \times B_R)$, and has no folds, that is, we suppose that $\det(\Phi_x(t, x)) \neq 0$ for (t, x) in $I \times B_R$.

We define the manifold

$$\mathbf{M} = \{(t, x, \Phi(t, x)): (t, x) \text{ in } I \times B_R\}$$

which we refer to as being the *quasistatic*, or reduced, *manifold*.

Next, we suppose that the quasi-static manifold is stable:

Hypothesis H5. The system of equations

$$\frac{dY}{d\tau} = g(\alpha, \beta, Y, 0)$$

has $Y = \Phi(\alpha, \beta)$ as an equilibrium for each (α, β) in $I \times B_R$. We suppose that this equilibrium is *asymptotically stable uniformly in the parameters* (α, β) in $I \times B_R$. That is, there is a comparison function $\mathbf{a}$, a positive monotone decreasing function $\mathbf{d}$ for which $\mathbf{d}(\infty) = 0$, and a positive constant δ such that if $|Y(\tau_0) - \Phi(\alpha, \beta)| < \delta$, then the corresponding solution, $Y(\tau, \alpha, \beta)$, exists and satisfies

$$|Y(\tau, \alpha, \beta) - \Phi(\alpha, \beta)| \leq \mathbf{a}(|Y(\tau_0) - \Phi(\alpha, \beta)|)\mathbf{d}(\tau - \tau_0)$$

for $0 < \tau_0 \leq \tau < \infty$.

With these conditions we have the following lemma:

Lemma. If conditions H3, H4, and H5 are satisfied, then there is a function $W(\alpha, \beta, Y)$ such that

1. W is twice continuously differentiable on $I \times B_R \times B_{R'}$.
2. $\mathbf{a}(|y - \Phi(\alpha, \beta)|) \leq W(\alpha, \beta, y) \leq \mathbf{b}(|y - \Phi(\alpha, \beta)|)$.
3. $g(\alpha, \beta, y, 0) \cdot \mathrm{grad}_y\, W(\alpha, \beta, y) \leq -\mathbf{c}(|y - \Phi(\alpha, \beta)|)$,

where **a**, **b**, and **c** are comparison functions that are independent of (α, β) in $I \times B_R$.

This lemma is proved in Ref. [8.1]. The proof is quite similar to that created by Massera [8.12] to construct Liapunov functions (see Chapter 3). W is a Liapunov function for the quasistatic manifold, and it serves as the basis for proving the following theorem:

Quasistatic Manifold Theorem. *Suppose that conditions H3, H4, and H5 are satisfied, and let the initial condition be such that η is in the domain of attraction of $\Phi(t_0, \xi_0)$ for the system in H5 with $\alpha = t_0$ and $\beta = \xi_0$. Then for each small ε, there is a unique solution of Eqs. (8.4) for $t_0 \le t \le T$. Moreover, the solution satisfies*

$$d(y, M) = o(1) \qquad as \ \varepsilon \to 0^+$$

uniformly on any interval of the form $t_0 < t_1 \le t \le T$, that is, one not including the initial time t_0. This result holds for any nearby (Caratheodory) system as well.

The result is not proved here (see 8.1). However, the main idea is to use the function W to define a "sticky" layer that encloses the set M. The derivative of W along a solution of Eqs. (8.4) is

$$\varepsilon \frac{dW}{dt} = \nabla W \cdot g + \varepsilon \frac{\partial W}{\partial t} + \varepsilon \nabla W \cdot f$$

The argument proceeds as it did for stability under persistent disturbances. Since the last two terms are $O(\varepsilon)$, there is a level set of W that is of diameter $o(1)$ (for small ε) and that defines an attractive and invariant layer about M.

Example. Lorenz's system of equations [8.35] comprises three equations for the amplitudes of dominant modes in the convective flow of a fluid in a toroidal container that is heated from below. These equations were extracted from among all possible modes by an assumption that the system of three amplitude equations is closed. In fact, the system is embedded in a system of higher dimension, which we illustrate here with the following system in E^4

$$\frac{dx_1}{dt} = x_2 x_3 - b x_1 + \varepsilon f_1(t, x, y, \varepsilon)$$

$$\frac{dx_2}{dt} = -x_1 x_3 + r x_3 - x_2 + \varepsilon f_2(t, x, y, \varepsilon)$$

$$\frac{dx_3}{dt} = \sigma(x_2 - x_3) + \varepsilon f_3(t, x, y, \varepsilon)$$

$$\frac{dy}{dt} = \lambda y - y^3 + \varepsilon g(t, x, y, \varepsilon)$$

where f_1, f_2, f_3, and g are Caratheodory functions. The quasistatic manifold $(y = O(\sqrt{\lambda}) + O(\varepsilon))$ can be constructed when λ and ε are near zero. When $\varepsilon = 0$, the system reduces to Lorenz's system, which is known to be chaotic for various choices of b, r, and σ. In this case, if λ is near 0, then $W(y) = y^2$, and the previous theorem shows that the solutions remain near the quasistatic manifold for all $t > 0$. Since the dynamics on the manifold are chaotic, we cannot expect the solution of the full problem to remain near any one particular solution of the reduced problem for fixed t as $\varepsilon \to 0^+$. The same result can be extended to vectors y, even infinite dimensional ones, using methods derived in Ref. 8.23. This completes the example.

Let (x_0, y_0) denote the solution of the reduced problem (8.5) with $y_0(t) = \Phi(t, x_0(t))$, and we have the following corollary

Corollary. *If the conditions of the previous theorem are satisfied, if $T < \infty$, and if the solution of the reduced problem $(x_0(t), y_0(t))$ exists for $t_0 \leq t \leq T$, then on any interval of the form $t_0 < t_1 \leq t \leq T$, for sufficiently small ε, the solution of the full problem (8.4), $(x(t), y(t))$, exists and it satisfies*

$$x(t) = x_0(t) + o(1), \qquad y(t) = \Phi(t, x_0(t)) + o(1)$$

uniformly as $\varepsilon \to 0^+$.

The proof of this result follows from the observation that the theorem gets this solution near $(x_0(t_1), y_0(t_1))$, and solutions depend continuously on the data to leading order in ε [see Ref. 8.1 for details].

We next place stability conditions on motions in the quasistatic manifold.

Hypothesis H6. Suppose now that the system (8.5) $dx_0/dt = f(t, x_0, \Phi(t, x_0), 0)$ has a solution for $t_0 \leq t < \infty$, say $x^*(t)$, and it is uniformly asymptotically stable. Furthermore, suppose that ξ_0 is in the domain of attraction of x^*.

With this additional condition, we have the following theorem

Quasistatic-State Theorem. *Let conditions H3, H4, H5, and H6 be satisfied. Then for sufficiently small values of ε, the solution of (8.4) exists for $t_0 \leq t < \infty$, and it satisfies*

$$(x, y) = (x_0(t), \Phi(t, x_0(t))) + o(1)$$

as $\varepsilon \to 0^+$, uniformly on any interval of the form $t_0 < t_1 \leq t < \infty$. Moreover, this result holds for any nearby (Caratheodory) system.

Sketch of Proof. Condition H6 implies that there is a Liapunov function $V(t, x)$ for which

1. $V(t, x^*(t)) = 0$.
2. There are comparison functions $\mathbf{a}_1$ and $\mathbf{b}_1$ such that

$$\mathbf{a}_1(|x - x^*|) \le V(t, x) \le \mathbf{b}_1(|x - x^*|)$$

for x near $x^*(t)$.

3. There is a comparison function $\mathbf{c}_1$ such that

$$\frac{\partial V}{\partial t} + \nabla V(t, x) \cdot f(t, x, \Phi(t, x), 0) \le -\mathbf{c}_1(|x - x^*(t)|)$$

for $0 \le t < \infty$ and x near $x^*(t)$.

The derivative of V along solutions of (8.4) is

$$\frac{dV}{dt} = \frac{\partial V}{\partial t} + \nabla V(t, x) \cdot f(t, x, y, \varepsilon)$$

If $y = \Phi(t, x) + o(1)$ (as from the previous theorem), then

$$\frac{dV}{dt} \le -\mathbf{c}_1(|x - x^*(t)|) + o(1)$$

Proceeding as for stability under persistent disturbances in Section 3.4, we see that there is a level set of V, having diameter $o(1)$ as $\varepsilon \to 0^+$, which is attractive and invariant. This combined with the Quasistatic Manifold Theorem shows that there is a "sticky tube" about $(x_0(t), \Phi(t, x_0(t)))$. Details of these proofs are given in Ref. 8.1. ∎

As an example of this result, we consider the system

$$\frac{dx}{dt} = -x^3 + \varepsilon f(t, x, y, \varepsilon)$$

$$\varepsilon \frac{dy}{dt} = -y^3 + \varepsilon g(t, x, y, \varepsilon)$$

where f and g are Caratheodory functions. $V = x^2$ and $W = y^2$ are the desired Liapunov functions, and so the result applies to this system:

$$(x(t), y(t)) = (0, 0) + o(1)$$

for each $t > 0$.

8.3.2   Matched Asymptotic Expansions

We turn now to approximating the solutions of system (8.4). For this we require a condition that is similar to invertibility in the implicit function theorem.

Denote by $g_y(t)$ the Jacobian matrix

$$g_y(t) = \left[\frac{\partial g_i}{\partial y_j}(t, x_0(t), y_0(t), 0) \right]$$

for $i, j = 1, \cdots, n$.

Hypothesis H7. Suppose that system (8.5) has a smooth solution $(x_0(t), y_0(t))$ and all of the eigenvalues of the matrix

$$g_y(t)$$

satisfy Re $\lambda(t) \leq -\delta < 0$, on some finite interval $0 \leq t \leq T$.

Next, we suppose that the data are smooth near the solution of the reduced problem.

Hypothesis H8. The functions f and g have continuous derivatives up to order $N + 2$ with respect to their arguments in some neighborhood of the points $(t, x_0(t), y_0(t))$, $0 \leq t \leq T$, and for $0 \leq \varepsilon < \varepsilon_0$. Also, the initial data $(\xi(\varepsilon), \eta(\varepsilon))$ are smooth functions of ε for $0 \leq \varepsilon < \varepsilon_0$.

With these conditions, we have the main theorem of this section:

Quasistatic-State Approximation Theorem. *Let conditions H3, H4, H7, and H8 be satisfied. Then there is a neighborhood U of $(x_0(0), y_0(0))$ such that for each small $\varepsilon > 0$ the problem (8.4) has a unique solution for $0 \leq t \leq T$ provided $(\xi, \eta) \in U$. Moreover, there is a quasistatic state $x^*(t, \varepsilon)$, $y^*(t, \varepsilon)$ and an initial transient $X(\tau, \varepsilon)$, $Y(\tau, \varepsilon)$ such that*

$$x(t, \varepsilon) = x^*(t, \varepsilon) + X\left(\frac{t}{\varepsilon}, \varepsilon\right)$$

$$y(t, \varepsilon) = y^*(t, \varepsilon) + Y\left(\frac{t}{\varepsilon}, \varepsilon\right)$$

for $0 \leq t \leq T$ and each small $\varepsilon > 0$. The quasistatic state has the form

$$x^*(t, \varepsilon) = \sum_{j=0}^{N} x_j(t)\varepsilon^j + O(\varepsilon^{N+1})$$

$$y^*(t, \varepsilon) = \sum_{j=0}^{N} y_j(t)\varepsilon^j + O(\varepsilon^{N+1})$$

where the error estimates hold uniformly for $0 \leq t \leq T$, and the initial transient has the form

$$X\left(\frac{t}{\varepsilon}, \varepsilon\right) = \sum_{j=0}^{N} X_j\left(\frac{t}{\varepsilon}\right)\varepsilon^j + O(\varepsilon^{N+1})$$

$$Y\left(\frac{t}{\varepsilon}, \varepsilon\right) = \sum_{j=0}^{N} Y_j\left(\frac{t}{\varepsilon}\right)\varepsilon^j + O(\varepsilon^{N+1})$$

where the error estimates hold uniformly for $0 \leq t \leq T$ as $\varepsilon \to 0$. Finally, there are positive constants K, α, and ε' such that

$$|X(\tau, \varepsilon)| + |Y(\tau, \varepsilon)| \leq K |\eta(\varepsilon) - y^*(0, \varepsilon)| \exp(-\alpha\tau)$$

for $0 \leq \tau \leq T/\varepsilon$ and $0 < \varepsilon \leq \varepsilon'$.

The proof of this result is presented in Ref. 8.2, and it is not presented here. However, the first steps in the proof require constructing the Taylor expansions listed in the theorem, and this is done next. Once these have been derived, the proof is completed by obtaining integral equations for the remainders which are the differences between the solution of the full problem and the proposed approximations, and then showing that they have unique solutions that are of order $O(\varepsilon^{N+1})$. This proof was carried out in detail for linear systems in Section 8.2.

Vasil'eva [8.9, 8.13] developed a method of matched asymptotic expansions to solve problem (8.4), and it is from her work that the present approach grew. Her algorithm is described in Refs. 8.9 and 8.13. Other refinements and extensions of her method were carried out by her students and colleagues [see 8.2]. Roughly, their approach entails the construction of three expansions, an outer one, an inner one, and a third one used to match these two. The inner and matching expansions are combined here in the initial transient of the QSS method. The matching approach has proved to be quite useful in studying a variety of problems, and Refs. 8.3 and 8.5 describe many features of the method and its uses in fluid mechanics and elasticity.

8.3.3 *Construction of QSSA*

The coefficients in the expansion of the quasistatic state are found by solving problem (8.5) for x_0 and y_0, and then for $j = 1, \cdots, N$ by solving

$$\frac{dx_j}{dt} = f_x x_j + f_y y_j + p_j(t), \qquad x_j(0) = \xi_j^*$$

$$\frac{dy_{j-1}}{dt} = g_x x_j + g_y y_j + q_j(t)$$

where for each j the functions p_j and q_j depend on the coefficients x_k, y_k, for $k = 0, \cdots, j - 1$. The initial condition for x_j (i.e., ξ_j^*) is determined at a later step by the matching conditions. These N problems have unique solutions.

The expansion of the initial transient is determined by solving for $j = 0$

$$\frac{dX_0}{d\tau} = 0, \qquad\qquad X_0(0) = \xi_0 - \xi_0^*$$

$$\frac{dY_0}{d\tau} = g(0, X_0, Y_0, 0), \qquad Y_0(0) = \eta_0 - y^*(0,0)$$

and for $j = 1, \cdots, N$, by solving

$$\frac{dX_j}{d\tau} = P_j(\tau), \qquad\qquad X_j(0) = \xi_0 - \xi_0^*$$

$$\frac{dY_j}{d\tau} = g_x X_j + g_y Y_j + Q_j(\tau), \qquad Y_j(0) = \eta_0 - y_j^*(0)$$

The functions P_j and Q_j are determined by the previous coefficients X_k, Y_k, for $k = 1, \cdots, j - 1$.

The values of ξ_j^* are determined at each step to ensure that the matching conditions $X_j(\infty) = 0$, $Y_j(\infty) = 0$, are satisfied. These are given by the formulas

$$\xi_j^* = \int_0^\infty P_j(\tau') \, d\tau' + \xi_j$$

Each of these N problems has a unique solution. In this way, the expansions in the theorem are found.

8.3.4   The Case $T = \infty$

The results of the QSSA Theorem are valid over the entire half-line $0 \le t < \infty$ if the reduced problem satisfies additional stability conditions. We now assume the following.

Hypothesis H9. The solution of the reduced problem (x_0, y_0) exists and remains bounded for all $0 \le t < \infty$, and the linear problem

$$\frac{dx}{dt} = [f_x(t) - f_y(t)g_y^{-1}(t)g_x(t)]x$$

has $x = 0$ as an exponentially asymptotically stable solution.

If condition H9 is added to the hypotheses of the QSSA Theorem with $T = \infty$, then the results of that theorem are valid for $0 \le t < \infty$. In particular, the quasistatic state and the initial transient can be expanded in powers of ε as indicated there, and the error estimates hold as $\varepsilon \to 0^+$ uniformly for $0 \le t < \infty$. This extension of the QSSA Theorem is proved in Ref. 8.2.

Example. The Michaelis-Menten Approximation. A simple enzyme reaction involves an enzyme E, a substrate S, a complex C, and a product P. Such reactions were studied by Michaelis and Menton and by Haldane and Briggs [8.14] Their work is widely used in biochemistry. Schematically, the reaction is

$$E + S \rightleftharpoons C \rightleftharpoons E + P$$

The forward reactions dominate the backward ones. After some preliminary scaling to remove dimensions from the problem, this reaction can be described by a system of differential equations for the (normalized) substrate concentration (x) and the (normalized) complex concentration (y) as

$$\frac{dx}{dt} = -x + (x + k)y, \qquad x(0) = 1$$

$$\varepsilon \frac{dy}{dt} = x - (x + K)y \qquad y(0) = 0$$

where ε measures a typical ratio of enzyme to substrate concentration $(O(10^{-5}))$ and k and K ($k < K$) denote normalized (nondimensional) rate constants.

The QSS method shows that the reduced problem is

$$\frac{dx_0}{dt} = \frac{(k - K)x_0}{(K + x_0)}$$

with the initial condition $x_0(0) = 1$, and

$$y_0 = x_0/(K + x_0)$$

The rapid transient is determined from the equation

$$\frac{dY_0}{d\tau} = -(1 + K)Y_0, \qquad Y_0(0) = -\frac{1}{1 + K}$$

Moreover, since the equation for x_0 is exponentially stable, we conclude that the approximation

$$x = x_0(t) + O(\varepsilon)$$

$$y = y_0(t) + Y_0(t/\varepsilon) + O(\varepsilon)$$

is valid uniformly for $0 \le t < \infty$.

This approximation is known as the Michaelis-Menten approximation. The equation

$$\frac{dx_0}{dt} = \frac{(k - K)x_0}{K + x_0}$$

gives the quasistatic approximation to this reaction, and, while the rate equations are based on the law of mass action, this equation seems to involve a more complicated rate law. $V_{\max} = K - k$ is called the *uptake velocity* or maximum reaction rate, and K is the reaction's *saturation constant*. These two constants are used by biochemists to characterize enzyme action.

8.4   Singular Perturbations of Oscillations

Let us now consider the time-invariant problem

$$\frac{dx}{dt} = f(x, y, \varepsilon)$$

$$\varepsilon \frac{dy}{dt} = g(x, y, \varepsilon)$$

A particularly interesting problem arises when the reduced problem

$$\frac{dx_0}{dt} = f(x_0, y_0, 0)$$

$$0 = g(x_0, y_0, 0)$$

has a stable oscillation. Two interesting questions in this case are whether or not the full problem has a stable oscillation near this reduced one, and whether or not the period of this solution is near the reduced oscillation's period.

It is useful to keep an example in mind, say

$$\frac{dx_1}{dt} = f_1(x, y, \varepsilon)$$

$$\frac{dx_2}{dt} = f_2(x, y, \varepsilon)$$

$$\varepsilon \frac{dy}{dt} = g(x, y, \varepsilon)$$

in E^3 where g has the form of a cusp:

$$g(x, y, \varepsilon) = -y^3 + x_1 y - x_2 + \varepsilon G(x, y, \varepsilon)$$

We see that when $\varepsilon = 0$ the solutions of the reduced problem lie on a cusp surface as shown in Figure 8.4.

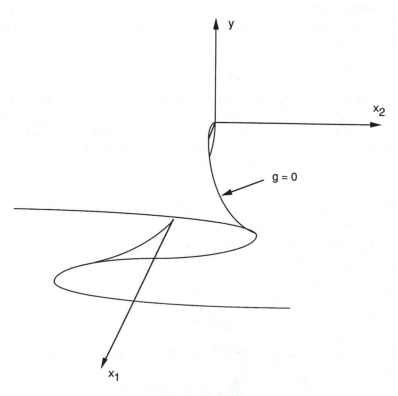

FIGURE 8.4. Cusp.

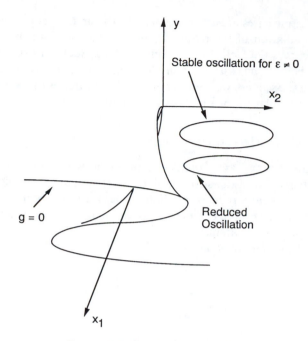

FIGURE 8.5. Quasistatic oscillation.

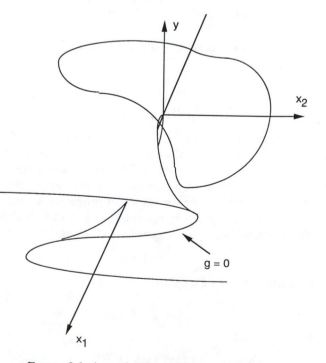

FIGURE 8.6. A nearly discontinuous oscillation.

Two interesting cases are described here. In the first, we suppose that there is a stable oscillation of the reduced problem that does not hit one of the bifurcation curves (i.e., folds) of the cusp surface. Such an oscillation is referred to as being a *quasistatic oscillation*, and a typical case is depicted in Figure 8.5. In the second, we consider an oscillation that is draped over a fold in the surface, similar to the one shown in Figure 8.6.

8.4.1   *Quasistatic Oscillations*

There is an interesting class of oscillation problems where a periodic solution can be constructed as a quasistatic solution. In the cusp example, these are solutions that lie near a reduced oscillation that is on one branch of the cusp surface, as shown in Figure 8.5. Because of this, computation of approximate solutions is quite transparent.

Consider the system of equations

$$\frac{dx}{dt} = f(x, y, \varepsilon)$$

$$\varepsilon \frac{dy}{dt} = g(x, y, \varepsilon)$$

where $x, f \in E^m$ and $y, g \in E^n$. We suppose that f and g are smooth functions of their variables in some region of E^{m+n}, and we restrict attention to that region.

The results of this section are based on the reduced problem

$$\frac{dx}{dt} = f(x, y, 0)$$

$$0 = g(x, y, 0)$$

having a periodic solution, say (x_0, y_0), and having period T_0. Conditions are found that ensure the existence of a nearby periodic solution of the full problem and that it can be constructed. The construction involves only a quasistatic solution of the problem, and so it is quite straightforward.

As we have seen with other time-invariant problems, we must expect the perturbed problem to have a periodic solution of a slightly different period than T_0. Therefore, we introduce an unknown time scaling

$$t = T(\varepsilon)s$$

into the problem with the result that

$$\frac{dx}{ds} = T(\varepsilon)f(x, y, \varepsilon)$$

$$\varepsilon \frac{dy}{ds} = T(\varepsilon)g(x, y, \varepsilon)$$

If the quasistatic solution is periodic, we can construct the oscillation in the form

$$x = x_0(s) + x_1(s)\varepsilon + x_2(s)\varepsilon^2 + \cdots$$

$$y = y_0(s) + y_1(s)\varepsilon + y_2(s)\varepsilon^2 + \cdots$$

In addition to finding the coefficients in these expansions, we must also show how the t-period $T(\varepsilon)$ is to be chosen. The equations for the coefficients are

$$\frac{dx_0}{ds} = T_0 f(x_0, y_0, 0)$$

$$0 = T_0 g(x_0, y_0, 0)$$

all of which are known, and for $j = 1, 2, \cdots$,

$$\frac{dx_j}{ds} = T_j \frac{dx_0}{ds} + T_0(f_x x_j + f_y y_j) + p_j$$

$$\frac{dy_{j-1}}{ds} = T_0(g_x x_j + g_y y_j) + q_j$$

where the functions p_j and q_j depend on x_k, y_k, and T_k for $k = 0, \cdots, j - 1$. Here the functions g_x, and so on, denote Jacobian matrices of derivatives, $(\partial g_i/\partial x_k)(x_0, y_0, 0)$, and so forth.

Hypothesis H10. Suppose the reduced problem has a periodic solution, x_0, $y_0 = \Phi(t, x_0(t))$, say with period T_0, and the matrix $g_y(x, y, 0)$ is stable for (x, y) near (x_0, y_0).

With this assumption, we can solve the second equation for y_j, and so the problem reduces to one for x_j:

$$\frac{dx_j}{ds} = T_j \frac{dx_0}{ds} + A(s)x_j + P_j(s)$$

where P_j is a one-periodic function of s that depends on terms with index k, $k < j$, and the matrix A is periodic of period 1 that is given by the formula

$$A(s) = T_0(f_x - f_y g_y^{-1} g_x)$$

Note that (dx_0/ds) is a periodic solution of the linear problem

$$du/ds = A(s)u$$

Therefore, there is a deficiency in the number of independent state variables that is made up for by the presence of the period's coefficients, similar to the situation in studying stability and regular perturbations of periodic solutions to autonomous systems (see Sections 3.5.4. and 6.1.2).

Hypothesis H11. Suppose the linear problem

$$du/ds = A(s)u$$

has exactly one characteristic exponent of the form $2\pi i N$ for some integer N.

With this, as before, there is a unique choice of T_j such that each of the problems for $x_1, \cdots, x_N$ has a periodic solution of period T_0. There remains a free constant in x_j that correspond to the initial phase, but that can be fixed by taking the first component of x_j to be zero. With such a condition, x_j is uniquely determined. This calculation is justified by the following theorem.

Friedrichs-Wasow Theorem. *Suppose that conditions* H10 *and* H11 *are satisfied and that the functions* f *and* g *are smooth near* (x_0, y_0). *Then for sufficiently small* $\varepsilon > 0$, *there is a unique periodic solution of the full problem with period* $T(\varepsilon)$, *that lies near* (x_0, y_0). *Moreover,* $T(\varepsilon) = T_0 + O(\varepsilon)$. *This solution is orbitally asymptotically stable if* (x_0, y_0) *is also.*

The proof of this result follows directly from the construction of the periodic quasistatic state. Again, remainders are defined and integral equations are derived for them. The proof is not presented here [see 8.10].

Example of the Friedrichs-Wasow Theorem. The example studied in this section uses some results from our earlier work on averaging. Consider the system in E^3

$$\frac{du}{dt} = -\lambda v + \mu\left(u - \frac{u^3}{3}\right) + \varepsilon f(u, v, y, \varepsilon, \mu)$$

$$\frac{dv}{dt} = \lambda u + \varepsilon F(u, v, y, \varepsilon, \mu)$$

$$\varepsilon\frac{dy}{dt} = -Dy + g(u, v) + \varepsilon G(u, v, y)$$

where u and v are scalars and y, g, $G \in E^n$. λ is a fixed frequency, and μ is a fixed small (positive) number in this system. We will consider the case where

$$0 < \varepsilon \ll \mu \ll 1.$$

Setting $\varepsilon = 0$ in this system gives

$$\frac{du_0}{dt} = -\lambda v_0 + \mu\left(u_0 - \frac{u_0^3}{3}\right)$$

$$\frac{dv_0}{dt} = \lambda u_0$$

$$Dy_0 = g(u_0, v_0)$$

We see that the first two equations reduce to van der Pol's equation, which we studied in Sections 2.1.3 and 7.4.2. In particular,

$$u_0(t) = r(\mu t)\cos \lambda t + O(\mu)$$

$$v_0(t) = r(\mu t)\sin \lambda t + O(\mu)$$

where $r(\infty) = 2$. The third equation is easily solved if D is invertible. In fact, let us suppose that $-D$ is a stable matrix (i.e., all of its eigenvalues have negative real parts). Then

$$y_0(t) = D^{-1}g(r(\mu t)\cos \lambda t + O(\mu), r(\mu t)\sin \lambda t + O(\mu))$$

This reduced solution is orbitally asymptotically stable since solutions approach the quasistatic manifold at an exponential rate determined by the eigenvalues of D and solutions in the manifold approach van der Pol's oscillation.

A periodic quasistatic state can be constructed for this system, but rather than proceeding with this result, we simply apply the Friedrichs-Wasow Theorem and obtain an easy approximation to this oscillation by using the first terms of the quasistatic state:

$$u = 2\cos(\lambda t) + O(\mu) + O(\varepsilon)$$

$$v = 2\sin(\lambda t) + O(\mu) + O(\varepsilon)$$

$$y = D^{-1}g(2\cos(\lambda t), 2\sin(\lambda t)) + O(\mu) + O(\varepsilon)$$

If we wish to include the rapid transients in this approximation, we use the quasistatic state approximation and write

$$u(t, \varepsilon) = r(\mu t)\cos(\lambda t) + O(\mu) + O(\varepsilon)$$

$$v(t, \varepsilon) = r(\mu t)\sin(\lambda t) + O(\mu) + O(\varepsilon)$$

$$y(t, \varepsilon) = D^{-1}g(r(\mu t)\cos(\lambda t), r(\mu t)\sin(\lambda t)) + \exp(-Dt/\varepsilon)[y(0) - y_0(0)] + O(\mu)$$
$$+ O(\varepsilon)$$

In this way, we quickly derive an approximation to the solution of a quite complicated system of equations.

It is interesting to pursue a slightly different approach to the problem— one that enables us to calculate the period of the oscillation. Let us return to the original problem

$$\frac{du}{dt} = -\lambda v + \mu\left(u - \frac{u^3}{3}\right) + \varepsilon f(u, v, y, \varepsilon, \mu)$$

$$\frac{dv}{dt} = \lambda u + \varepsilon F(u, v, y, \varepsilon, \mu)$$

$$\varepsilon\frac{dy}{dt} = -Dy + g(u, v) + \varepsilon G(u, v, y)$$

and let us introduce polar coordinates in the first two equations: Let

$$u = \rho \cos \phi$$
$$v = \rho \sin \phi$$

This results in the new system

$$\frac{d\rho}{dt} = \mu\rho \cos^2 \phi \, (1 - \rho^2 \cos^2 \phi) + O(\varepsilon)$$

$$\frac{d\phi}{dt} = \lambda - \mu \sin \phi \cos \phi \, (1 - \rho^2 \cos^2 \phi) + O(\varepsilon)$$

$$\varepsilon \frac{dy}{dt} = -Dy + g(\rho \cos \phi, \rho \sin \phi, y) + \varepsilon G(\rho \cos \phi, \rho \sin \phi, y)$$

Since ϕ is a timelike variable, we can write

$$\frac{d\rho}{d\phi} = \mu\rho \cos^2 \phi \frac{1 - \rho^2 \cos^2 \phi}{\lambda} + O(\mu^2) + O(\varepsilon)$$

$$\varepsilon \frac{dy}{d\phi} = \frac{-Dy + g(\rho \cos \phi, \rho \sin \phi, y)}{\lambda} + O(\mu^2) + O(\varepsilon)$$

The QSS Theorem can be applied directly to this system to construct its quasistatic state. Once this is done, we return to compute the period of the quasistatic state by solving the equation

$$\frac{d\phi}{\lambda - \mu \sin(\phi) \cos(\phi)(1 - \rho^2 \cos^2(\phi)) + O(\varepsilon)} = dt$$

In particular, we have that

$$dt = \frac{d\phi}{\lambda + O(\mu, \varepsilon)}$$

and so

$$T(\varepsilon, \mu) = \frac{2\pi}{\lambda} + O(\mu, \varepsilon)$$

The higher order corrections to the period can be calculated directly by using the corresponding corrections in the integral defining T.

This last calculation clears up the confusion about having one too few state variables that often results. When a time-invariant system is converted to phase-amplitude coordinates, the natural time scale for solutions is the new phase variable, not t, and the period T results from converting from the period in the phase variable, for example 2π in the case of polar coordinates, to t.

8.4.2 *Nearly Discontinuous Oscillations*

Let us now consider briefly the case where the system

$$\frac{dx}{dt} = f(x, y, 0)$$

$$0 = g(x, y, 0)$$

has a periodic solution along which condition H1 is not satisfied. Specifically, we suppose that

Hypothesis H12. The reduced problem has a discontinuous solution of period T_0, and $\det(g_y(x_0(t), y_0(t), 0)) \neq 0$ for all but a finite set of t values. We suppose that $g_y(t, x_0(t), y_0(t), 0)$ is stable away from these points.

The problem now is similar to that depicted in Figure 8.6, and it is significantly more difficult than the preceding case.

Oscillations of this kind were studied by van der Pol, Cartwright, and Littlewood [8.15], Levinson [8.16], Pontryagin [8.17], Mishchenko [8.18], Kevorkian and Cole [8.6], Stoker [8.7], and Grasman [8.19], among others. They are often called *relaxation oscillations*. Unfortunately, there is no result comparable to the Friedrichs-Wasow Theorem to ensure the existence of stable oscillations in this case, except for special cases including the ones cited above.

The t values at which g_y vanishes are called *turning points*, and there is an extensive theory for constructing solutions near them [see 8.5, 8.10, 8.17, 8.18]. Rather than pursuing a vague general discussion of these problems, we turn to a specific, well-studied example.

Van der Pol's Equation

Van der Pol's equation has been studied extensively, so it provides a useful guide to solving other nearly discontinuous oscillation problems. Consider the equation

$$\varepsilon \frac{d^2x}{dt^2} + (x^2 - 1)\frac{dx}{dt} + x = 0$$

where ε is a small positive parameter. It is useful to rewrite this equation as a first-order system of equations by integrating it once. The result is

$$\varepsilon \frac{dx}{dt} = x - \frac{x^3}{3} - y$$

$$\frac{dy}{dt} = x$$

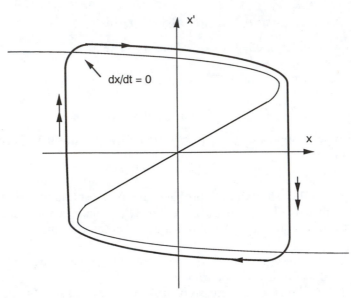

FIGURE 8.7. Depiction of van der Pol's relaxation oscillation, $\varepsilon \ll 1$. This shrink-wraps onto the isocline as $\varepsilon \to 0^+$.

Four questions about this equation that have been answered are as follows:

1. Does this equation have a periodic solution for small values of ε?
2. If so, what is its period?
3. How can such an oscillation be constructed?
4. What is the response of this oscillator to external periodic forcing?

The first question is answered by our work in Section 3.5 on Lienard's equation. There is a unique periodic solution to this equation for any choice of the parameter ε, and it is globally asymptotically stable.

The phase-portrait of this solution is shown in Figure 8.7 where $\varepsilon \ll 1$. The isocline $dx/dt = 0$ has been drawn in this portrait for reference. For small ε, the oscillation remains quite close to the cubic x isocline except at the jump (or turning) points. The jumps are traversed quite rapidly, in a time of order ε (double arrow).

The period can be approximated by observing that most of the time is spent near the cubic curve. This is described by the reduced problem for this system:

$$\frac{dx_0}{dt} = \frac{x_0}{(1 - x_0^2)}$$

and the part of this solution lying near half of the oscillation occurs for $1 < x_0 < 2$. Therefore, the transit time of this branch is given (approximately) by

$$T_0 = 2 \int_2^1 \frac{1 - x^2}{x} \, dx = 3 - 2\ln 2$$

It follows that the period is given by

$$T \sim 3 - 2\ln 2 + o(1)$$

as $\varepsilon \to 0$. In fact, further work (that is not presented here [see 8.10]) shows that

$$T = 3 - 2\ln 2 + 7.014\varepsilon^{2/3} - 0.167\varepsilon \ln \varepsilon - 1.325\varepsilon + O(\ln \varepsilon / \varepsilon)$$

The third question is more difficult to answer, and it is not addressed here, but the details of the construction can be found in Ref. 8.5. Finally, the response of this oscillator to external periodic forcing can be chaotic. This is described in Ref. 8.20 where further references can be found [see also 8.15, 8.16, 8.19].

8.5 Boundary-Value Problems

Singularly perturbed boundary-value problems arise in many important applications. As we have seen, finding periodic solutions entails solving a problem with boundary conditions (e.g., $x(0) = x(T)$). Models for spatial structure on whose boundaries physical parameters are set, such as temperature, pressure, electrical current, and so on, and many control problems are boundary-value problems [see, e.g., 8.3]. An understanding of the role played by ε in these problems is helped by understanding the geometric structure of solutions. Rather than developing a complete boundary-layer theory here, we study an interesting example that gives the flavor of how the QSS analysis can be used to solve boundary-value problems.

A Josephson junction is a cryogenic device consisting of two superconductors separated by a thin gap [8.21]. The electrodynamics of this device are described in terms of a quantum mechanical wave function that is quite similar to the phase variable used to describe VCOs. A jump occurs in the wave function across the gap, and it is denoted by $u(x, t)$, where x, $0 \le x \le 1$, describes the location along the gap and t is time. The sine-Gordon equation is satisfied by u:

$$\frac{\partial^2 u}{\partial t^2} + \sigma \frac{\partial u}{\partial t} - \varepsilon^2 \frac{\partial^2 u}{\partial x^2} + \sin u = 0$$

for $0 \le x \le 1$ and $t \ge 0$.

A magnetic field of strength proportional to H is applied at both ends, and the device is driven by a current I applied at the right end. These conditions are described by the boundary conditions

$$\frac{\partial u}{\partial x}(0, t) = H, \qquad \frac{\partial u}{\partial x}(1, t) = H + I$$

The static states of this configuration are found by solving the problem

$$\varepsilon^2 \frac{d^2 u}{dx^2} - \sin u = 0$$

with the boundary conditions

$$\frac{du}{dx}(0) = H, \qquad \frac{du}{dx}(1) = H + I$$

We consider here only the case where $\varepsilon \ll 1$.

We rewrite this problem as a first-order system of equations

$$\varepsilon \frac{du}{dx} = v$$

$$\varepsilon \frac{dv}{dx} = \sin u$$

where now $v(0) = \varepsilon H$ and $v(1) = \varepsilon(H + I)$. We must construct a solution of this system that lies on the line $v = \varepsilon H$ at $x = 0$ and meets the line $v = \varepsilon(H + I)$ when $x = 1$.

A candidate for a quasistatic solution is $u = 0$, $v = 0$. We test this by trying to constuct correcting (transients) at both end points ($x = 0$ and $x = 1$) that connect the boundary conditions to this quasistatic state. We seek a *left boundary-layer correction* $u = \varepsilon U(x/\varepsilon)$, $v = \varepsilon V(x/\varepsilon)$ such that

$$V(0) = H \qquad \text{and} \qquad (U(\xi), V(\xi)) \to (0,0) \qquad \text{as } \xi \to \infty$$

The problem for (U, V) is

$$\frac{dU}{d\xi} = V$$

$$\frac{dV}{d\xi} = \frac{1}{\varepsilon} \sin \varepsilon U = U + O(\varepsilon)$$

Figure 8.8 shows the phase portrait of solutions.

U and V can be constructed in power series in ε: $U = U_0 + \varepsilon U_1$, and so on. We see that there is a unique solution for (U_0, V_0) that satisfies these two conditions: It is

$$U_0(\xi) = -He^{-\xi}, \qquad V_0(\xi) = He^{-\xi}$$

This starts at the intersection of the stable manifold for $(0,0)$ and the constraint $V = H$.

Next, we construct the right boundary-layer correction. Let $\eta = (1 - x)/\varepsilon$. Then $\eta = 0$ corresponds to $x = 1$ and $\eta \to +\infty$ corresponds to $(1 - x)/\varepsilon \to \infty$, so η represents a fast scale that moves backward from $x = 1$. The problem for $u = \varepsilon U(\eta)$ and $v = \varepsilon V(\eta)$ is

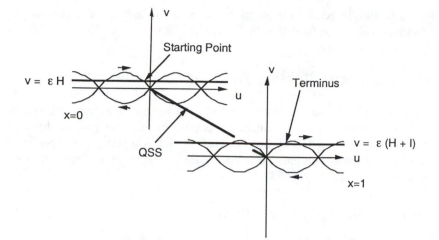

FIGURE 8.8. Geometry of solutions to the boundary-value problem.

$$\frac{dU}{d\eta} = -V$$

$$\frac{dV}{d\eta} = -\frac{1}{\varepsilon}\sin \varepsilon U = -U + O(\varepsilon)$$

$$V(0) = H + I \quad \text{and} \quad (U(\eta), V(\eta)) \to (0, 0) \quad \text{as } \eta \to \infty.$$

This can be constructed in a power series in ε. As before, we see that there is a unique solution for $U_0(\eta)$, $V_0(\eta)$:

$$U_0(\eta) = (H + I)e^{-\eta}, \qquad V_0(\eta) = (H + I)e^{-\eta}$$

Combining these results, we might expect that an approximate solution of the problem is

$$u = -He^{-x/\varepsilon} + (H + I)e^{(x-1)/\varepsilon} + O(\varepsilon)$$

$$v = \varepsilon(He^{-x/\varepsilon} + (H + I)e^{(x-1)/\varepsilon} + O(\varepsilon))$$

which is the sum of a left boundary-layer correction, a right boundary-layer correction, and a quasistatic state. There remains an important question: Is there a solution to this boundary-value problem that lies near (within $O(\varepsilon)$) the approximation that we constructed? The answer lies in a detailed analysis of trajectories of the equation. If ε is so small that $\varepsilon(H + I) < 1$, then there is such a solution. The proof is constructed as in other problems: An integral equation is derived for the remainder, that is the difference between (u, v) and the approximation, and that equation is shown to have a solution [see 8.3].

There are many kinds of boundary-value problems involving singular perturbations that must be studied. The one described here illustrates the basic geometry involved in many of these problems. However, a survey, such as

Ref. 8.22 should be consulted to get a better feeling for the diversity of boundary-value problems that have been studied.

8.6 Nonlinear Stability Analysis near Bifurcations

The singular perturbation methods described in this chapter are closely related to stability results, as we have seen repeatedly. Because of this, it is not surprising that a QSSA can be used to establish stability properties of various problems. This has been particularly important in problems from fluid mechanics [see Ref. 8.23].

8.6.1 *Bifurcating Static States*

The QSSA method is particularly useful in cases where changing parameter values results in the appearance of a new static state through bifurcation. Consider the system

$$dz/dt = F(z, \lambda)$$

where z, $F \in E^N$ and λ is a real parameter. Suppose that there is a static state, say $z = 0$, which loses its stability at some value, say $\lambda = \lambda_0$. In particular, assume the following condition.

Hypothesis H13. $F(0, \lambda) = 0$ for all λ near $\lambda = \lambda_0$. The Jacobian matrix $F_z(0, \lambda)$ is assumed to be stable for $\lambda < \lambda_0$, it has a single eigenvalue equal to zero for $\lambda = \lambda_0$, and it has a single eigenvalue with positive real part for $\lambda > \lambda_0$. All the other eigenvalues are assumed to have negative real parts for λ near λ_0.

With assumption H13, we can determine whether or not a new static state appears for the problem as λ increases through λ_0 by using the Liapunov-Schmidt method described in Chapter 4.

Let ϕ denote the null eigenvector of $F_z(0, \lambda_0)$ and write

$$z = c\phi + w$$

where $w \cdot \phi = 0$. The static state problem

$$F(z, \lambda) = 0$$

becomes

$$0 = PF(c\phi + w, \lambda)$$

$$0 = Bw + QF(c\phi + w, \lambda)$$

where P is the projection of E^N onto ϕ and Q is the complementary projection. The matrix B is a stable matrix found by projecting the Jacobian matrix $F_z(0, 0)$ onto the complement of ϕ.

As shown in Chapter 4, there is a unique solution for w, say $w = w^*(c, \lambda)$ for c near zero and λ near λ_0, and substituting this into the first equation gives the bifurcation equation

$$0 = PF(c\phi + w^*(c, \lambda), \lambda)$$

If this equation has nontrivial small solutions for c as functions of $\lambda - \lambda_0$, then new static states appear through a simple bifurcation.

A general theory for using the QSSA to test the stability of these new static states is presented in Ref. 8.23. We consider here an example that illustrates typical conditions when the method is useful.

Example. Landau's Equation for a Canonical Bifurcation Problem. The problem described in Section 4.3.1 is amenable to nonlinear stability analysis. Consider the system of equations

$$\frac{dx}{dt} = \lambda x - axy + \text{h.o.t.}$$

$$\frac{dy}{dt} = -by + dx^2 + \text{h.o.t.}$$

Here a, b, and d are fixed positive numbers, and λ is a parameter. We saw that as λ increases through zero ($\lambda_0 = 0$ here), a pair of new static states appears given by

$$(x, y) = \left(\pm \sqrt{\frac{b\lambda}{ad}}, \frac{\lambda}{a} \right) + \text{h.o.t.}$$

The stability of these new states is determined in Section 4.3.1 by linearizing the problem about them and finding that these linear problems are stable. The Linear Stability Theorem then applies to show that these are stable solutions of the nonlinear problem.

One deficiency in using linear stability methods is that they are restricted to a neighborhood of the static state, possibly to an extent where they do not accurately describe the domain of attraction of the state. The following approach based on the QSSA can correct this.

Newton's polygon analysis in Section 4.1 indicates the appropriate scaling to use in this problem. For $\lambda > 0$, we set $\varepsilon = \lambda$, $x = \sqrt{\varepsilon}\, u$, $y = \varepsilon v$, and $s = \varepsilon t$. With these changes of variables, the problem becomes

$$\frac{du}{ds} = u - auv + \text{h.o.t.}$$

$$\varepsilon \frac{dv}{ds} = -bv + du^2 + \text{h.o.t.}$$

Given *any* initial data $(u(0), v(0))$, we can determine the behavior of solutions of this system for ε near zero by using the QSSA Theorem. The reduced problem is

$$\frac{du_0}{ds} = u_0 - au_0v_0$$

$$0 = -bv_0 + du_0^2$$

There is a unique solution for v_0, and substituting this into the first equation gives

$$\frac{du_0}{ds} = u_0\left(1 - \frac{ad}{b}u_0^2\right), \qquad u_0(0) = u(0)$$

The solution of this problem exists for all $0 \le s < \infty$, and it approaches

$$\mathrm{sgn}(u(0))\sqrt{\frac{b}{ad}}$$

as $s \to \infty$ at an exponential rate.

The initial transient is determined by the equation

$$\frac{dY_0}{ds} = -bY_0, \qquad Y_0(0) = v(0) - \frac{d}{b}u(0)^2$$

Clearly, condition H5 of the QSSA Theorem is satisfied, and we conclude that

$$u = u_0(s) + O(\varepsilon)$$

$$v = v_0(s) + Y_0(s/\varepsilon) + O(\varepsilon)$$

where the error estimate holds uniformly for $0 \le s \le \infty$.

Returning to the original question, we can now state that if $x(0) = O(\sqrt{\varepsilon})$ and $y(0) = O(\varepsilon)$, say $x(0) = \sqrt{\varepsilon}\,\xi$ and $y(0) = \varepsilon\eta$ where ξ and η are fixed constants, then

$$x(t) \to \mathrm{sgn}(\xi)\sqrt{\frac{b\varepsilon}{ad}} + O(\varepsilon)$$

as $t \to \infty$.

The essential difference between the nonlinear stability analysis and the linear stability analysis is that the stability of the quasistatic state is determined by solving a nonlinear problem. This gives a more realistic description of the domain of attraction of the bifurcated state. Here there is no further restriction of $(x(0), y(0))$ other than $x(0) = O(\sqrt{\varepsilon})$ and $y(0) = O(\varepsilon)$, so the nonlinear stability analysis establishes that the unstable static state is in the boundary of the domain of attraction of the new static state. The equation for u_0 is referred to as *Landau's equation* in applications to fluid dynamics [8.24].

8.6.2 *Nonlinear Stability Analysis of Nonlinear Oscillations*

The Hopf theory of bifurcations that was discussed briefly in Chapter 4 can be studied for stability by the methods of the preceding section. In fact, consider the problem

$$dx/dt = f(x, \lambda)$$

which we suppose has a static state, say $x = \phi(\lambda)$. Suppose that for $\lambda < \lambda_0$, all of the eigenvalues of the Jacobian matrix

$$f_x(\phi(\lambda), \lambda)$$

lie in the left half of the complex plane, but that there are two eigenvalues, say $\rho(\lambda) \pm i\omega(\lambda)$ for which $\rho(\lambda_0) = 0$, $\rho'(\lambda_0) > 0$, and $\omega(\lambda_0) > 0$. The other eigenvalues are assumed to remain in the left half-plane for λ near λ_0. In this situation, which was described in Sections 4.3 and 6.1.4, a *Hopf bifurcation* might occur from the static state $x = \phi(\lambda)$ resulting in the appearance of a new oscillation. The stability analysis of this new oscillation can proceed in the following steps.

1. Project the problem onto the two modes carrying the purely imaginary eigenvalues, and their complement, all of which are damped modes.
2. Determine the manifold of quasiequilibrium values for the damped modes. Substitute this into the equations for the two oscillatory modes, which reduces it to a system for two variables.
3. Since the problem carried by the oscillatory modes is described by a perturbation of a harmonic oscillator, introduce polar coordinates into these equations.
4. Using the phase variable of this system as a timelike variable, reduce the system to a single scalar equation.
5. Apply the Mean-Stable Averaging Theorem to this equation to derive the analog of Landau's equation.

For example, consider the problem

$$x'' + \varepsilon x' + \omega^2 x = \varepsilon f(x, y, x', y', \varepsilon)$$

$$y'' + ry' + \mu^2 y = \varepsilon g(x, y, x', y', \varepsilon)$$

where f and g are smooth functions, and r, ω, and μ are fixed positive constants. For $\varepsilon = 0$, there results a linear problem whose spectrum consists of the four values

$$\pm i\omega, \qquad -r \pm \sqrt{r^2 - \mu^2}$$

Obviously, the y components will equilibrate on a fast time scale (relative to εt), so we set $y' = y'' = 0$. The result is that the quasistatic manifold is defined by

$$y_0 = 0$$

and so on. Thus, to leading order in ε, we have

$$x'' + \omega^2 x = \varepsilon f(x, 0, x', 0, 0) - \varepsilon x'$$

Introducing polar coordinates into this system using the formula

$$\frac{dx}{dt} + i\omega x = Re^{i\theta}$$

gives

$$\frac{d\theta}{dt} = \omega - \varepsilon \sin \theta \frac{f - R \cos \theta}{R}$$

where $f = f[R \sin(\theta)/\omega, 0, R \cos(\theta), 0, 0]$. Using θ as a timelike variable, we have

$$\frac{dR}{d\theta} = \frac{\varepsilon \cos \theta \, (f - R \cos \theta)}{\omega - \varepsilon \sin \theta (f/R - \cos \theta)} = \frac{\varepsilon}{\omega} \cos \theta (f - R \sin \theta) + O(\varepsilon^2)$$

Averaging this equation gives Landau's equation for the problem

$$\frac{dR}{d\theta} = -\frac{\varepsilon}{\omega} \left(\frac{1}{2\pi} \int_0^{2\pi} \cos T \, f\left(\frac{R}{\omega} \sin T, 0, R \cos T, 0, 0 \right) dT - \frac{R}{2} \right)$$

Static states of this equation correspond to periodic solutions of the original system, and, if such a static state is asymptotically stable for the averaged equation, then the oscillation is orbitally asymptotically stable (i.e., stable in amplitudes).

This nonlinear stability analysis of bifurcating oscillations uses both quasi-static state and averaging methods. The solutions reduce to a quasistatic manifold, and the highly oscillatory behavior on it can be described using averaging methods. The reader should carry out the details of this calculations for the case of van der Pol's equation where

$$f(x, 0, x', 0, \varepsilon) = x^2 x'$$

8.7   Explosion Mode Analysis of Rapid Chemical Reactions

Explosive chemical reactions are difficult to study using the QSSA because most of the interesting dynamics occur in transients after an initiation phase, but before the reaction comes to equilibrium. The example treated in this section illustrates a case where the QSSA theorem does not give useful information directly, but it provides some guidance in obtaining a *canonical problem* whose solution is relatively simple to find. The canonical problem gives an approximate description of the initial transient's behavior, where most of the interesting dynamics occur.

The combustion of hydrogen involves a chain-branched reaction that is described in Table 8.1 [8.25]. Three types of reactions are involved in this process: initiation, branching (or propagation), and termination. In these reactions, HO_2 is taken to be an inactive particle, and W in the termination reactions indicates collisions with the container's wall.

We describe the concentrations of various chemical species by

$$u = [H_2], \qquad v = [O_2], \qquad x = [H^{\cdot}], \qquad y = [OH^{\cdot}], \qquad z = [O^{\cdot}]$$

Then the kinetic rate equations can be derived directly from Table 8.1 as being

TABLE 8.1. H_2—O_2 elementary reactions.

Reaction[a]	Rate	Type of Reaction
$H_2 + O_2 \rightarrow H\cdot + HO_2$	k_0	Initiation
$H_2 + OH\cdot \rightarrow H\cdot + H_2O$	k_1	
$O_2 + H\cdot \rightarrow OH\cdot + O\cdot$	k_2	
$H_2 + O\cdot \rightarrow OH\cdot + H\cdot$	k_3	Propagation
$OH\cdot + W \rightarrow$	a_1	Termination
$H\cdot + W \rightarrow$	a_2	Termination
$O\cdot + W \rightarrow$	a_3	Termination

[a] $H\cdot$, $O\cdot$, etc. denote highly reactive radicals.

$$\frac{du}{dt} = -(k_1 y + k_3 z)u - k_0 uv \qquad\qquad u(0) = u_0$$

$$\frac{dv}{dt} = -k_2 xv - k_0 uv \qquad\qquad v(0) = v_0$$

$$\frac{dx}{dt} = -(k_2 v + a_1)x + k_1 uy + k_3 uz + k_0 uv \qquad x(0) = 0$$

$$\frac{dy}{dt} = k_2 vx - (k_1 u + a_2)y + k_3 uz \qquad\qquad y(0) = 0$$

$$\frac{dz}{dt} = k_2 vx - (k_3 u + a_3)z \qquad\qquad z(0) = 0$$

It is convenient to rewrite this system using matrices: Let

$$B = \begin{bmatrix} -k_2 v & k_1 u & k_3 u \\ k_2 v & -k_1 u & k_3 u \\ k_2 v & 0 & -k_3 u \end{bmatrix}$$

and

$$T = \begin{bmatrix} a_1 & 0 & 0 \\ 0 & a_2 & 0 \\ 0 & 0 & a_3 \end{bmatrix}$$

The matrix B describes the branching reactions, and the matrix T describes the termination reactions. If $X = \mathrm{col}(x, y, z)$, then

$$\frac{dX}{dt} = (B - T)X + \mathrm{col}(k_0 uv, 0, 0)$$

where the terms on the right-hand side denote branching, termination, and initiation, respectively, of the radical concentrations.

Spectral Analysis of $B - T$

The characteristic polynomial of $B - T$ is

$$P(\lambda) = -\det(B - T - \lambda I_3)$$

where I_3 is the 3×3-identity matrix. The roots of this polynomial are the eigenvalues of $B - T$. P has the form

$$P(\lambda) = \lambda^3 + \sigma\lambda^2 + v\lambda - \omega(u, v)$$

where $\sigma = \alpha + \beta + \kappa + a_1 + a_2 + a_3 > 0$,

$$v = \alpha(a_2 + a_3) + \beta(\kappa + a_1 + a_3) + \kappa(a_1 + a_2) + a_1 a_2 + a_3 a_2 + a_1 a_3 > 0$$

and

$$\omega(u, v) = 2\alpha\beta\kappa - [a_1 \beta\kappa + a_1 a_2 \beta + a_3 a_2 \alpha + a_1 a_2 a_3]$$

where $\alpha = k_2 v$, $\beta = k_1 u$, and $\kappa = k_3 u$. Note that

$$\omega(u, v) = \det(B - T)$$

The sign of ω plays an important role here. First, P is a monotone increasing function of λ since its first derivative is positive for $\lambda > 0$.

Therefore, if $P(0) = -\omega(u, v) < 0$, then there is a unique real, positive eigenvalue $\lambda^*(u, v)$. Moreover, if λ_1 and λ_2 denote the other two eigenvalues, then either (1) they are real and both negative, or (2) they are imaginary. In the last case,

$$\lambda^* + \lambda_2 + \lambda_1 = -\sigma < 0, \text{ so } 2\,\mathrm{Re}\,\lambda_1 < -\lambda^* < 0.$$

Thus, in either case, if $\omega(u, v) > 0$, then $B - T$ has one positive, real eigenvalue and two other eigenvalues that have negative real parts.

Denote by $\lambda^*(u, v)$ the eigenvalue of $B - T$ that has the largest real part, and let ϕ^* denote the corresponding eigenvector, so $(B - T)\phi^* = \lambda^*\phi^*$. ψ^* denotes the adjoint eigenvector

$$(B - T)^{\mathrm{tr}}\psi^* = \lambda^*\psi^*$$

These are normalized so that

$$\psi^* \cdot \phi^* = 1 \quad \text{and} \quad \psi^* \cdot \psi^* = 1.$$

ϕ^* is referred to as the *explosion mode* and λ^* gives its *amplification* or *explosion rate*. For relevant values of the reaction rates, ϕ^* is observed to remain essentially constant throughout the reaction.

The radical components can be rewritten as

$$\begin{pmatrix} x \\ y \\ z \end{pmatrix} = c\phi^* + \Omega$$

where the vector Ω accounts for the λ_1 and λ_2 modes. Since these modes will be damped, we ignore them and introduce the *explosion mode approximation*

$$\begin{bmatrix} x \\ y \\ z \end{bmatrix} = c\phi^*$$

Substituting this into the model gives

$$\frac{du}{dt} = -(k_1\phi_2^* + k_3\phi_3^*)\,cu \qquad -k_0 uv$$

$$\frac{dv}{dt} = -(k_2\phi_1^*)cv \qquad\qquad -k_0 uv$$

$$\frac{dc}{dt} = \lambda^*(u,v)c \qquad\qquad\quad +\alpha k_0 uv$$

where α gives the projection of the initiation vector $col(1,0,0)$ onto ϕ^*. Since the initiation reaction is negligible throughout most of the reaction, we ignore it:

$$\frac{du}{dt} = -(k_1\phi_2^* + k_3\phi_3^*)cu$$

$$\frac{dv}{dt} = -(k_2\phi_1^*)cv$$

$$\frac{dc}{dt} = \lambda^*(u,v)c$$

This is called the *canonical first-order branching problem*, and it is solvable; the equation for du/dv can be easily solved:

$$v = Ku^\xi, \qquad \text{where} \qquad \xi = \frac{(k_2\phi_1^*)}{(k_1\phi_2^* + k_3\phi_3^*)},$$

Then the equation for dc/du can be integrated:

$$c = \bar{c} - \int_{\bar{u}}^{u} \frac{\lambda^*(v, Kv^\xi)}{(k_1\phi_2^* + k_3\phi_3^*)} \frac{dv}{v}$$

where $\bar{u}$ and $\bar{c}$ are values after the initiation phase ($\bar{u} \sim u_0, \bar{c} \sim 0$).

As the reaction proceeds, u and v are depleted and $\lambda^*(u,v)$ moves from being large and positive to being negative. The values of u and v for which $\lambda^* = 0$ define the *first explosion limit* of the reaction. If u and v are supercritical ($\lambda^* > 0$), then an explosion will ensue that begins termination only when $\lambda^* = 0$. The values of c at termination can be estimated using the above formula for c [8.25, 8.26].

8.8   Computational Schemes Based on QSSA

The numerical solution of singular perturbation problems can be difficult for several reasons; for example, it is not unusual in applications that $\varepsilon = O(10^{-10})$. Solving for the initial transient can be handled by taking a very small step size for the numerical algorithm to ensure accuracy, but it is only the first problem. The second problem is that any numerical algorithm constantly makes errors that throw the solution off of the reduced solution, and small step sizes might be required throughout the calculation to ensure that the numerical scheme is stable. This last aspect of the computation is referred to as *stiffness* of the system. Various computer packages have been designed to circumvent this problem of stiffness, such as Gear' package and LSODE, but computation remains expensive [see 8.27, 8.28].

QSS methods can be used to formulate a useful numerical scheme that takes advantage of solution structure to avoid some of the stiffness problems. These can significantly reduce the computation time, but they do require preprocessing the system into a standard form which may not be possible in reasonable time.

Consider the initial value problem

$$\frac{dx}{dt} = f(t, x, y, \varepsilon), \qquad x(0) = \xi(\varepsilon) = \xi_0 + O(\varepsilon)$$

$$\varepsilon\frac{dy}{dt} = g(t, x, y, \varepsilon), \qquad y(0) = \eta(\varepsilon) = \eta_0 + O(\varepsilon)$$

where ε is a small, positive parameter. Here $x, f, \xi \in E^m$ and $y, g, \eta \in E^n$. The reduced problem ($\varepsilon = 0$ with the initial y condition canceled) is

$$\frac{dx}{dt} = f(t, x, y, 0), \qquad x(0) = \xi_0$$

$$0 = g(t, x, y, 0)$$

It is assumed to have a smooth solution, $x = x_0(t)$, $y = y_0(t)$, on some interval $0 \le t \le T$. Moreover, the data f, g, ξ, and η are assumed to be smooth near this solution and for ε near zero. Finally, the Jacobian matrix

$$g_y(t, x_0(t), y_0(t), 0)$$

is assumed to be stable (i.e., all of its eigenvalues lie in the left half-plane, bounded uniformly away from the imaginary axis for all $0 \le t \le T$).

Under these conditions, the Quasistatic-State Approximation Theorem shows that the solution of the full problem has the form

$$x(t, \varepsilon) = x_0(t) + x_1(t)\varepsilon + X(t/\varepsilon, \varepsilon) + O(\varepsilon^2)$$

$$y(t, \varepsilon) = y_0(t) + y_1(t)\varepsilon + Y(t/\varepsilon, \varepsilon) + O(\varepsilon^2)$$

where X and Y satisfy

$$|X(t/\varepsilon, \varepsilon)| + |Y(t/\varepsilon, \varepsilon)| \le Ke^{-\mu t/\varepsilon}$$

for some positive constants K and μ that are independent of ε. These estimates hold uniformly for $0 \le t \le T$.

We wish to determine a numerical approximation of $(x(h, \varepsilon), y(h, \varepsilon))$ for a given step size $h \gg \varepsilon$. Direct evaluation of this quantity by numerical packages can be expensive if ε is very small, but the QSSA Theorem ensures that $x_0(h), y_0(h)$ is a useful approximation.

8.8.1 Direct Calculation of $x_0(h), y_0(h)$

Since $h/\varepsilon \gg 1$, we can ignore X and Y terms in the approximation and $x_0(h)$, $y_0(h)$ should give an acceptable approximation to $x(h, \varepsilon), y(h, \varepsilon)$. We begin with a guess for $y_0(0)$ and use Newton's method to solve the equation

$$0 = g(0, x(0), y, 0)$$

for y. This process can be repeated as required up to $t = h$. Let us next use a pth-order numerical method for solving the ordinary differential equation for $x_0(h)$. The result is that

$$x(h, \varepsilon) = x_0(h) + O(\varepsilon) + O(h^{p+1})$$

$$y(h, \varepsilon) = y_0(h) + O(\varepsilon) + O(h^{p+1})$$

where $x_0(h)$ and $y_0(h)$ are the computed solutions.

8.8.2 Extrapolation Method

It is possible to avoid solving the reduced problem by taking advantage of all of the information given by the QSSA Theorem. The extrapolation method described here was derived in Ref. 8.29. The idea is to identify a value of ε, say ε', that is substantially larger than ε, but for which the solution $x(h, \varepsilon'), y(h, \varepsilon')$ approximates $x(h, \varepsilon), y(h, \varepsilon)$ to the accuracy of the numerical scheme used. Solving the full problem with a larger value of ε can be done with greater accuracy using less effort.

First, a value T is found such that

$$K \exp(-\mu T) = O(h^{p+1})$$

μ is usually of order of the largest negative eigenvalue of g_y, and K depends on size of g_y. Next, the value

$$\varepsilon' = h/T$$

is defined, and the full problem is solved using a standard integration method, say of order p, to determine values $x(h, \varepsilon'/2), y(h, \varepsilon'/2)$ and $x(h, \varepsilon')$, $y(h, \varepsilon')$. It follows that

$$x(h, \varepsilon) = 2x(h, \varepsilon'/2) - x(h, \varepsilon') + O(h^{p+1}) + O(\varepsilon'^2) + O(\varepsilon)$$

$$y(h, \varepsilon) = 2y(h, \varepsilon'/2) - y(h, \varepsilon') + O(h^{p+1}) + O(\varepsilon'^2) + O(\varepsilon)$$

These formulas are derived by observing that

$$2x(h, \varepsilon'/2) = 2x_0(h) + x_1(h)\varepsilon' + 2X(T, \varepsilon'/2) + O(\varepsilon'^2)$$

and

$$x(h, \varepsilon') = x_0(h) + x_1(h)\varepsilon' + X(T, \varepsilon') + O(\varepsilon'^2)$$

Subtracting these two expressions gives

$$2x(h, \varepsilon'/2) - x(h, \varepsilon') = x_0(h) + O(h^{p+1}) + O(\varepsilon'^2)$$

On the other hand,

$$x(h, \varepsilon) = x_0(h) + O(\varepsilon)$$

so the desired result for $x(h, \varepsilon)$ is found. Similarly, for $y(h, \varepsilon)$. The final result is that

$$x(h, \varepsilon) = 2x(h, \varepsilon'/2) - x(h, \varepsilon') + O(h^{p+1}) + O(\varepsilon) + O((h/T)^2)$$
$$y(h, \varepsilon) = 2y(h, \varepsilon'/2) - y(h, \varepsilon') + O(h^{p+1}) + O(\varepsilon) + O((h/T)^2)$$

This is referred to as the *extrapolation approximation*.

In many applications, the number of operations used in these computations is proportional to $1/\varepsilon$, for $x(h, \varepsilon)$, and so on, and to $1/\varepsilon'$, for $x(h, \varepsilon')$, and so on. Therefore, the ratio ε'/ε indicates the relative number of operations of direct solution compared to the extrapolation solution.

Using the extrapolation approximation can avoid analytical preprocessing of a problem to get it into the form of the full problem for x and y if K and μ can be estimated. It also avoids having to solve the algebraic equation $g = 0$ at each mesh point required for solving the x equation of the reduced problem. Both preprocessing and solving for the quasistatic state y can require substantial amounts of time, and a major object of solving packages is to avoid preprocessing.

This extrapolation method is based on the QSS perturbation scheme. It is distinct from straightforward Richardson extrapolation formulas that are based on Taylor's formula and that are not appropriate for stiff problems. It is important to note that this method improves as ε gets smaller.

8.9   Exercises

8.1. Use the quasistatic state approximation to construct the solution of the system

$$\frac{dx}{dt} = ax + by + \sin t \qquad x(0) = 1$$

$$\varepsilon \frac{dy}{dt} = -y + \cos t \qquad y(0) = 3$$

where x and y are scalars and ε is a small positive number. Construct both the quasistatic state and the initial layer corrections through order ε^2.

8.2. Show in the Quasistatic State Approximation Theorem that the initial layer corrections satisfy $X_j \to 0$ as $\tau \to \infty$.

8.3*. Problems involving two, or more, small parameters can also be handled using the QSS method. Consider the system

$$\frac{dx}{dt} = Ax + Bu + Cv + f(t)$$

$$\varepsilon \frac{du}{dt} = Du + Ev + g(t)$$

$$\varepsilon^2 \frac{dv}{dt} = Fv + h(t)$$

where x, f are in E^m, u, g are in E^n, and v, h are in E^k. The matrices A, B, and so on, are of appropriate dimensions. This problem contains two small parameters ε and ε^2. Suppose that the matrices D and F are stable and that the functions f, g, and h are smooth functions.

Show that this problem has a quasistatic state (x^*, u^*, v^*) and two transients

$$\left(X\left(\frac{t}{\varepsilon}, \varepsilon\right), U\left(\frac{t}{\varepsilon}, \varepsilon\right), V\left(\frac{t}{\varepsilon}, \varepsilon\right) \right)$$

and

$$\left(X^*\left(\frac{t}{\varepsilon^2}, \varepsilon\right), U^*\left(\frac{t}{\varepsilon^2}, \varepsilon\right), V^*\left(\frac{t}{\varepsilon^2}, \varepsilon\right) \right)$$

that decay at exponential rates. Moreover, show that the solution is given by the formula

$$(x, u, v) = (x^*, u^*, v^*) + (X, U, V) + (X^*, U^*, V^*)$$

and that the terms in this formula can be expanded in Taylor expansion about $\varepsilon = 0$.

8.4. Suppose that the functions $f(x, y, \varepsilon)$ and $g(x, y, \varepsilon)$ are smooth functions (at least twice continuously differentiable) for all real numbers x, y, and ε. Suppose that the system

$$\frac{dx_1}{dt} = f_1(x, y, 0)$$

$$\frac{dx_2}{dt} = f_2(x, y, 0)$$

$$0 = -y^3 + x_1 y - x_2$$

has an oscillation, say $x_1^*(t)$, $x_2^*(t)$, $y(t)$ on which

$$-3y^2 + x_1 < 0.$$

Therefore, the oscillation lies either on the top or bottom branch of the cusp surface. Let T_0 denote the period of this oscillation. The Friedrichs-Wasow Theorem shows that there is a unique periodic solution of the full problem lying near the orbit of the reduced problem and having period near T_0. Use the method of matched asymptotic expansions to derive the first-order approximation to the oscillation of the full system.

8.5. Consider the atoll oscillator which defines a flow on a torus having interesting periodic solutions:

$$\frac{dx}{dt} = 5.0(1.05 + \cos x - \cos_+ y)$$

$$\frac{dy}{dt} = 0.04(1.0 + \cos y + 10.0\cos_+ x)$$

Prove that there is a periodic solution to this system. Simulate the solution of this system by calculating the solution and plotting the results on a toroidal patch (x and y modulo 2π) as shown in the figure below.

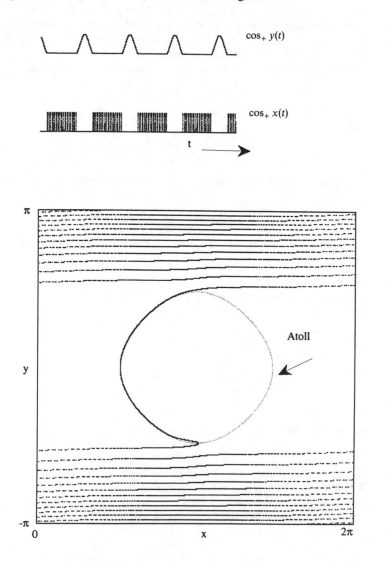

8.6. Consider the nonlinear stability problem

$$\frac{dx}{dt} = A(\lambda)x + f(t, x)$$

where x, f in E^m and A is a matrix of appropriate dimensions. Suppose that f is a smooth function that satisfies

$$f(t, x) = o(|x|)$$

as $|x| \to 0$ uniformly for $0 \le t < \infty$. Suppose that a simple eigenvalue of A passes through the origin from left to right as λ increases through zero while the remaining eigenvalues remain in the left-half plane, bounded away from the imaginary axis.

Apply the nonlinear stability analysis described in Section 8.6.1. Derive Landau's equation for this problem.

8.7. Consider the damped conservative system

$$\varepsilon u'' + ru' + U_u(u) = 0$$

where ε is a small dimensionless parameter, r describes the coefficient of friction and $U(u)$ describes the system's potential energy. We wish to solve this equation subject to the boundary conditions

$$u(0) = A, \qquad u(1) = B$$

Setting $\varepsilon = 0$ in the equation results in the reduced problem

$$ru_0' + U_u(u_0) = 0$$

This first-order differential equation's solution is uniquely determined by specifying u_0 at one point. Therefore, the solution cannot satisfy both boundary conditions (unless $A = B$).

a. What is the correct value for u_0?

b. Solve the problem explicitly in the special case where $U(u) = u^2/2$.

8.8. Show that if the matrix A in the quasistatic state theorem is exponentially stable, then the error estimates in the result hold uniformly for $0 \le t < \infty$.

8.9*. Consider the stochastic differential equation [8.30, 8.31]

$$dx = \frac{1}{\varepsilon} F'(x)\, dt + \sigma\, dW$$

where $x(t)$ is a random variable, and dW denotes identically distributed independent increments taken from a gaussian distribution having mean zero and variance 1. If $u(x, t)$ denotes the probability density function of the random variable x, then it is known that u solves the Fokker-Planck equation

$$\frac{\partial u}{\partial t} = \sigma^2 \frac{\partial^2 u}{\partial x^2} - \frac{\partial}{\partial x}\left(F'(x)\frac{u}{\varepsilon} \right)$$

a. Find a static state solution of this. (*Hint*: set $\partial u/\partial t = 0$ and solve the resulting equation for $u(x)$.)

b. Determine what happens to this distribution as ε and $\sigma^2 \to 0^+$. In particular, explain the role played by the maxima of $F(x)$ in the distribution of u.

8.10. Let p denote the proportion of a human population's gene pool that is carried by a single locus that is of type A, and let $q = 1 - p$ denote the proportion that are of type B. We suppose that there are only these two types. As a result, the population can be broken down into three groups AA, AB, and BB by their genetic type. Let $D(t)$, $2H(t)$, and $R(t)$ denote the proportions of the population of these types at time t. Then $p = (D + H)$, $q = (H + R)$. If selection is slow, then the birth rates and death rates for all of these genotypes are almost the same. Suppose that the birth rates are identical and that the death rates are $d_{AA} = d + \varepsilon d_1$, $d_{AB} = d + \varepsilon d_2$, $d_{BB} = d + \varepsilon d_3$. Then

$$\frac{dD}{dt} = b(D - p^2) + \varepsilon(d^* - d_1)D$$

$$\frac{dH}{dt} = b(H - pq) + \varepsilon(d^* - d_2)H$$

$$\frac{dR}{dt} = b(R - q^2) + \varepsilon(d^* - d_3)R$$

where $b > 0$ is the population's birth rate, and $d^* = d_1 D + 2d_2 H + d_3 R$ [8.32].

a. Derive a differential equation for p.
b. Apply the quasistatic state approximation theorem to this system to obtain the quasistatic state and the initial correction expansions to first order in ε.
c. Plot your results for $(D, 2H, R)$ using triangular coordinates, as shown in the figure below.

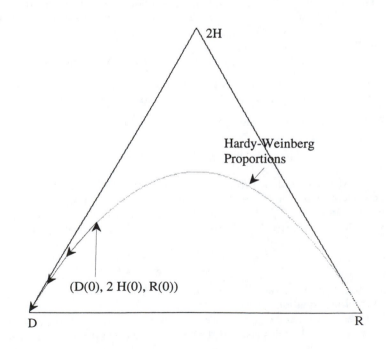

8.11*. Consider the nonlinear renewal equation [8.33]

$$\varepsilon x(t) = \int_{-1}^{0} \{(2\varepsilon - 1)x(t + s) - x^2(t + s)\}\, ds \qquad \text{for } t > 0$$

$$x(t) = \varepsilon v_0(t) \qquad \text{for } -1 \le t < 0$$

where $0 < \varepsilon \ll 1$ and where v_0 is extended as a periodic function. Use the multitime algorithm to find an approximate solution to this equation. In particular, let

$$x = \varepsilon V(x, t, \varepsilon) = \varepsilon V_0(\xi, \tau) + \varepsilon^2 V_1(\xi, \tau) + \cdots$$

where $\xi = (1 - \varepsilon)t$ and $\tau = \varepsilon^2 t$. Show that

$$\frac{\partial V_0}{\partial \tau} + \frac{\partial}{\partial \xi}(V_0 - V_0^2) = \frac{1}{2}\frac{\partial^2 V_0}{\partial \xi^2}$$

$$V_0(\xi, \tau) = V_0(\xi - 1, \tau) \qquad \text{and} \qquad V_0(\xi, 0) = v_0(\xi)$$

8.12*. Consider the equation [8.34]

$$\varepsilon \frac{\partial u}{\partial t} = D\frac{\partial^2 u}{\partial x^2} + u - u^3$$

for $0 \le x \le 1$ and the auxiliary boundary conditions

$$\frac{\partial u}{\partial x}(0, t) = 0, \frac{\partial u}{\partial x}(1, t) = 0$$

Suppose that $u(x, 0) = U(x)$ is given. Show that there is a quasistatic state for this problem. Construct it and the transient initial correction.

Supplementary Exercises

S.1. Consider the forced Duffing equation

$$x'' + \alpha(x - x^3) = B \cos 2\pi t$$

Calculate the rotation number of the response as a function of forcing amplitude B and oscillator tuning α by converting the problem to polar coordinates and calculating the values $\theta(100\pi)/100\pi$.

S.2. Solve the Michaelis-Menten model in Section 8.3.4 for $\varepsilon = 0.001$ and $\varepsilon = 0.00001$ using the extrapolation method described in section 8.8.

S.3. Consider a linear array of masses extending from $-\infty$ to $+\infty$, say having masses m_j at site j, $-\infty < j < \infty$. Suppose that the location of the jth mass is x_j, and that these masses are connected by springs. Then the dynamics of the ensemble are described by the equations

$$\frac{dx_j}{dt} = f(x_{j+1} - x_j) - f(x_j - x_{j-1})$$

where $f(u)$ describes the restoring force of a spring that is deflected u units from rest. Carry out a computer simulation of this system when $f(s) = as$, when $f(s) = bs + cs^3$, and when $f(s) = e^{-a|s|}$ where a, b, and c are fixed positive constants. The first case is a simple diffusion problem, the second is referred to as the Korteweg de Vries model, and the third is the Toda lattice.

S.4. Simulate the twist mapping

$$r_{n+1} = r_n, \qquad t_{n+1} = t_n + a[r]$$

where $[r]$ is the integer part of r.

S.5. Using a computer simulation, determine the two basins of attraction of the stable periodic solutions described in Figure 6.6. (*Hint*: Given a tolerance ε, calculate the probability that a line segment having this length and thrown at random hits the boundary. Calculate this number for several values of ε and extrapolate its value to $\varepsilon = 0$.)

S.6. Consider the array of VCONs

$$\frac{dx_j}{dt} = \omega_j + \varepsilon \tanh\left(A \cos x_j + C_j \cos_+ \mu t + \sum B_{j,k} \cos_+ x_k\right)$$

288

for $j = 1, \cdots, 100$, where the matrix B is a cyclic matrix. Take $\omega_j = 1.0 - 0.8*j/50.0$ for $j = 1, \cdots, 50$ and $\omega_j = \omega_{101-j}$ for $j = 51, \cdots, 100$, $A = 5.0$, $\varepsilon = 0.5$, $C_j = 5.0$, $B_{jk} = 0.0$.

a. Simulate this system and plot the resulting rotation vector $(\rho_1, \cdots, \rho_N)$ using polar coordinates. (Hint: Define $\rho_j = x_j(100\pi)/x_1(100\pi)$ and plot the points $(\rho_j \cos 2\pi j/100, \rho_j \sin 2\pi j/100)$ for $j = 1, \cdots, 100$.)

b. Experiment with various choices of the connection matrix B. For example, let it describe nearest neighbor interactions where it is a cyclic matrix having center row $\cdots 1-21 \cdots$, or let C be a companion matrix.

S.7. Consider the semiimplicit scheme

$$x_{n+1} = x_n - hf(y_{n+1})$$

$$y_{n+1} = y_n + hf(x_n)$$

where $f(u) = \sin u$. Simulate the solution of this iteration for various choices of initial conditions. Plot your answers by plotting the iterates of this mapping in the xy plane.

S.8. Carry out a Monte Carlo simulation of solutions to the stochastic differential equation

$$dy = \frac{1}{\varepsilon} F'(y) dt + \frac{\sigma}{\varepsilon} dW$$

by computing sample paths using the discrete model

$$y_{n+1} = y_n + \frac{h}{\varepsilon}(F'(y_n) + \sigma W_n)$$

where W_n is at each step a random variable selected from a normal distribution and F is a quartic function. Plot your answer as a histogram of y_{100}. (Hint: Use a random number generator for a uniformly distributed random variable on the interval $0 < y < 1$ to select six values. Add them up and divide by $\sqrt{6}$. The result is a normally distributed variable [see Hammersley and Handscombe].)

References

0.1. R. Thom, *Structural Stability and Morphogenesis: An Outline of a General Theory of Models*, W.A. Benjamin, Reading, Mass., 1975.

0.2. M.A. Cartwright, J.E. Littlewood, *Ann. Math.* 54(1951): 1–37.

0.3. A.A. Andronov, A.A. Witt, S.E. Chaikin, *Theory of Oscillators*, Dover, New York, 1966.

0.4. Salvador Dali, Dreamscape. Centre Georges Pompidou, Musee national d'art moderne, Paris, 1980.

0.5. E.A. Coddington, N. Levinson, *Theory of Ordinary Differential Equations*, McGraw-Hill, New York, 1955.

0.6. J.K. Hale, *Ordinary Differential Equations*, Wiley-Interscience, New York, 1971.

0.7. M. Hirsch, S. Smale, *Differential Equations, Dynamical Systems and Linear Algebra*, Academic Press, New York, 1974.

0.8. F.R. Gantmacher, *Applications of the Theory of Matrices*, Wiley-Interscience, New York, 1959.

0.9. H.S. Thompson, *Fear and Loathing in Las Vegas*, Fawcett, New York, 1971.

0.10. S. Freud, *The Interpretation of Dreams*, Allen & Unwin, London, 1954.

1.1. P. Horowitz, W. Hill, *The Art of Electronics*, 2nd ed., Cambridge University Press, New York, 1989.

1.2. R.J. Higgins, *Electronics with Digital and Analog Integrated Circuits*, Prentice-Hall, Englewood Cliffs, New Jersey, 1983.

1.3. R. Courant, D. Hilbert, *Methods of Mathematical Physics*, Vol I, Wiley-Interscience, New York, 1968.

1.4. J.S. Frame, *Applications of Matrices in Engineering*, MSU Lecture Notes, 1965.

1.5. D.K. Fadeev, V.N. Fadeeva, *Computational Methods in Linear Algebra*, W.H. Freeman, San Francisco, 1963.

1.6. F.R. Gantmacher, *Applications of the Theory of Matrices*, Wiley-Interscience, New York, 1959.

1.7. E.A. Coddington, N. Levinson, *The Theory of Ordinary Differential Equations*, McGraw-Hill, New York, 1955.

1.8. W. Magnus, S. Winkler, *Hill's Equation*, Wiley-Interscience, New York, 1966.

1.9. A.S. Besicovitch, *Almost Periodic Functions*, Dover, New York, 1954.

1.10. N. Wiener, *The Fourier Integral and Certain of its Applications*, Dover, New York, 1958.

2.1. E.A. Coddington, N. Levinson, *The Theory of Ordinary Differential Equations*, McGraw-Hill, New York, 1955.

2.2. J.K. Hale, *Ordinary Differential Equations*, Wiley-Interscience, New York, 1971.

2.3. M. Hirsch, S. Smale, *Differential Equations, Dynamical Systems and Linear Algebra*, Academic Press, New York, 1974.

2.4. A. Denjoy, Sur les courbes definies par les equations differentielles a la surface du tor, *J. Math. Pures Appl.* 9(1932): 333–375.

2.5. F.C. Hoppensteadt, *Introduction to the Mathematics of Neurons*, Cambridge University Press, New York, 1986.

2.6. P. Horowitz, W. Hill, *The Art of Electronics*, 2nd ed., Cambridge University Press, New York, 1989.

2.7. W. Chester, The forced oscillations of a simple pendulum, *J. Inst. Maths. Appl.*, 15(1975): 298–306.

2.8. M. Levi, F.C. Hoppensteadt, W.L. Miranker, Dynamics of the Josephson junction, *Quart. Appl. Math.* (July 1978): 167–198.

2.9. R. Courant, D. Hilbert, *Methods of Mathematical Physics*, Vol I, Wiley-Interscience, New York, 1968.

2.10. J. Smoller, *Shock Waves and Reaction-Diffusion Equations*, Springer-Verlag, New York, 1983.

2.11. F.C. Hoppensteadt, *Mathematical Methods of Population Biology*, Cambridge University Press, New York, 1982.

2.12. A.T. Fomenko, Integrable systems on Lie algebras and symmetric spaces, Gordon and Breach, New York, 1988.

2.13. M. Abramowitz, I.A. Stegun, *Handbook of Mathematical Functions*, Dover, New York, 1972.

2.14. A.I. Khinchin, *An Introduction to Information Theory*, Dover, New York, 1965.

2.15. C.L. Siegel, J. Moser, *Lectures in Celestial Mechanics*, Springer-Verlag, New York, 1971.

2.16. P.R. Garabedian, *Partial Differential Equations*, Wiley, New York, 1964.

2.17. *Preconditioned Conjugate Gradient Methods*, Springer-Verlag, New York, 1990.

2.18. R. Thom, *Structural Stability and Morphogenesis: An Outline of a General Theory of Models*, W.A. Benjamin, Reading, Mass., 1975.

2.19. A.N. Sarkovski, *Ukr. Math. Zh.* 16(1964): 61–71. See also P. Stefan, A theorem of Sarkovskii on the existence of periodic orbits of continuous endomorphisms of the real line, *Comm. Math. Phys.* 54(1977): 237–248.

2.20. T.Y. Li, J.A. Yorke, Period three implies chaos, *Amer. Math. Montly*, 82(1975): 985–992.

2.21. S. Ulam, *A Collection of Mathematical Problems*, Wiley-Interscience, New York, 1960.

2.22. M.E. Munroe, *Introduction to Measure and Integration*. Addison-Wesley, Cambridge, Mass., 1953.

2.23. F.C. Hoppensteadt, J.M. Hyman, Periodic solutions of a logistic difference equation, *SIAM J. Appl. Math.* 58(1977): 73–81.

2.24. K. Krohn, J.L. Rhodes, *Algebraic Theory of Machines* (M.A. Arbib, ed.), Academic Press, New York, 1968.

2.25. N.N. Minorsky, *Nonlinear Oscillations*, Van Nostrand, Princeton, 1962.

2.26. J. Moser, On the theory of quasi-periodic motions, *SIAM Rev.* 8(1966): 145–171.

2.27. G.D. Birkhoff, *Dynamical Systems*, Vol. IX., American Mathematics Society, Providence, RI, 1966.

2.28. J. Hadamard, Sur l'iteration et les solutions asymptotiques des equations différentielles, *Bull. Soc. Math. France* 29(1901): 224–228.

2.29. V.I. Arnol'd, A. Avez, Ergodic problems in classical mechanics, W.A. Benjamin, New York, 1968.

2.30. R. Bellman, K. Cooke, *Differential-Difference Equations*, Academic Press, New York, 1963.

2.31. P.E. Sobolevski, Equations of parabolic type in Banach space, *AMS Translation*, 49(1966): 1–62.

2.32. H.T. Banks, F. Kappel, Spline approximations for functional differential equations, *J. Differential Eqns.*, 34(1979): 496–522.

2.33. R.D. Nussbaum, H.O. Peitgen, Special and spurious solutions of $dx/dt = -aF(x(t-1))$, preprint.

3.1. W. Hahn, *Stability of Motion*, Springer-Verlag, New York, 1967.

3.2. E.A. Coddington, N. Levinson, *The Theory of Ordinary Differential Equations*, McGraw-Hill, New York, 1955.

3.3. H. Antosiewicz, A survey of Liapunov's second method, in *Contributions to the Theory of Nonlinear Oscillations*, S. Lefschetz (ed.), Vol. IV, Princeton, N.J., 1958.

3.4. J.K. Hale, *Ordinary Differential Equations*, Wiley-Interscience, New York, 1971.

3.5. W.A. Coppel, *Stability and Asymptotic Behavior of Differential Equations*, Heath, Boston, 1965.

3.6. P. Hartmann, *Ordinary Differential Equations*, Hartmann, Baltimore, 1973.

3.7. J.L. Massera, J.J. Schaffer, *Linear Differential Equations and Function Spaces*, Academic Press, New York, 1966.

3.8. O. Perron, Uber stabilitat und asymptotisches verhalten der Integrale von Differential-gleichgungenssysteme, *Math. Zeit.* 29(1929): 129–160.

3.9. M. Hirsch, S. Smale, *Differential Equations, Dynamical Systems and Linear Algebra*, Academic Press, New York, 1974.

3.10. *Preconditioned Conjugate Gradient Methods*, Springer-Verlag, New York, 1990.

3.11. F.R. Gantmacher, *Applications of the Theory of Matrices*, Wiley-Interscience, New York, 1959.

3.12. J.L. Massera, Contributions to stability theory, *Ann. Math.*, 64(1956): 182–206; 68(1958): 202.

3.13. I.G. Malkin, *Theorie der Stabilitat einer Bewegung*, Oldenbourg, Munich, 1959.

3.14. T. Yoshizawa, *Stability Theory and the Existence of Periodic and Almost Periodic Solutions*, Springer-Verlag, New York, 1975.

3.15. F.C. Hoppensteadt, *An Introduction to the Mathematics of Neurons*, Cambridge University Press, New York, 1986.

3.16. A. Denjoy, Sur les courbes definies par les equations differentielles a la surface du tor, *J. Math. Pures Appl.* 9(1932): 333–375.

3.17. D.E. Woodward, Phase locking in model neuron networks having group symmetries., Dissertation, University of Utah, 1988.

4.1. R. Courant, D. Hilbert, *Methods of Mathematical Physics*, Vol I, Wiley-Interscience, New York, 1968.

4.2. F.R. Gantmacher, *Applications of the Theory of Matrices*, Wiley-Interscience, New York, 1959.

4.3. E.W. Cheney, D. Kincaid, *Numerical Mathematics and Computing*, Brooks-Cole, Monterey, CA, 1980.

4.4. M.M. Vainberg, V.A. Trenogin, *Theory of Branching of Solutions of Nonlinear Equations*, Noordhoff, Leyden, 1974.

4.5. E. Hille, *Analytic Function Theory*, Vol. 1 and 2, Ginn, New York, 1959.

4.6. N. Gordon, F.C. Hoppensteadt, Nonlinear stability analysis of static states which arise through bifurcation, *Comm. Pure Appl. Math.* 28(1975): 355–373.

4.7. G. Iooss, D. Joseph, *Nonlinear Dynamics and Turbulence*, Pitman, Boston, 1983.

4.8. E. Hopf, *Ber. Math. Phys. Sachsische Akad. Wiss. Leipzig*, 94(1942): 1–22.

4.9. J. Dugundji, *Fixed Point Theory*, Panstwowe Wydawnictwo Naukowe, Warsaw, 1982.

4.10. L. Cesari, *Asymptotic Behavior and Stability Problems in Ordinary Differential Equations*, Ergebnisse der Math. New Series, Vol. 16, 1963, 2nd ed.

4.11. S.N. Chow, J.K. Hale, *Methods of Bifurcation Theory*, Springer-Verlag, New York, 1982.

4.12. R.D. Nussbaum, H.O. Peitgen, Special and spurious solutions of $dx/dt = -aF(x(t-1))$, preprint.

5.1. A. Erdelyi, *Asymptotic Expansions*, Dover, New York, 1956.

5.2. E.W. Cheney, *Introduction to Approximation Theory*, McGraw-Hill, New York, 1966.

5.3. E.A. Coddington, N. Levinson, *The Theory of Ordinary Differential Equations*, McGraw-Hill, New York, 1955.

5.4. K.O. Friedrichs, *Lectures on Advanced Ordinary Differential Equations*, Gordon and Breach New York, 1965.

6.1. J.B. Keller, *Perturbation Theory*, Michigan State University, 1968.

6.2. K.O. Friedrichs, *Lectures on Advanced Ordinary Differential Equations*, Gordon and Breach, New York, 1965.

6.3. G. Duffing, *Erzwungene Schwingungen Bie Veranderlicher Eigenfrequnz*, F. Viewg u. Sohn, Braunschweig, 1918.

6.4. C. Hayashi, *Nonlinear Oscillations in Physical Systems*, Princeton University Press, Princeton, 1985.

6.5. A.N. Sarkovski, *Ukr. Math. Zh.* 16(1964): 61–71. See also P. Stefan, A theorem of Sarkovskii on the existence of periodic orbits of continuous endomorphisms of the real line, *Comm. Math. Phys.* 54(1977): 237–248.

6.6. F.C. Hoppensteadt, J.M. Hyman, Periodic solutions of a logistic difference equation, *SIAM J. Appl. Math.* 58(1977): 73–81.

6.7. J.J. Stoker, *Nonlinear Vibrations*, Wiley-Interscience, New York, 1950.

6.8. M. Levi, On van der Pol's equation, *AMS Memoirs*, Providence, R.I., 1979.

6.9. P. Fatou, Sur les equations fonctionelles, *Bull Soc. Math. Fr.* 47(1919): 161–271; 48(1920): 33–94, 208–314.

6.10. G. Julia, Sur l'iteration des fonctions rationelles, *J. Math. Pure Appl.* 8(1918): 47–245.

6.11. J. Kinney, T. Pitcher, *Invariant Measures for Rational Functions*. Some connections between ergodic theory and the iteration of polynomials, Ark. Mat. 8(1969): 25–32.

6.12. B.B. Mandelbrot, *The Fractal Geometry of Nature*, Updated and Augmented. W.H. Freeman, New York, 1983.

6.13. J.E. Flaherty, F.C. Hoppensteadt, Frequency entrainment of a forced van der Pol oscillator, *Studies Appl. Math.* 58(1978): 5–15.

6.14. S.W. McDonald, C. Grebogi, E. Ott, J.A. Yorke, Fractal basin boundaries, *Physica* 17D (1985): 125–153.

6.15. J.D. Farmer, E. Ott, J.A. Yorke, *Physica* 7D (1983): 153.

6.16. M.E. Munroe, *Introduction to Measure and Integration*, Addison-Wesley, Cambridge, Mass., 1953.

7.1. W.A. Coppel, *Stability and Asymptotic Behavior of Differential Equations*, Heath, Boston, 1965.

7.2. J.K. Hale, *Ordinary Differential Equations*, Wiley-Interscience, New York, 1971.

7.3. F.C. Hoppensteadt, W.L. Miranker, Differential equations having rapidly changing solutions: Analytic methods for weakly nonlinear systems, *J. Diff. Eqn.* 22(1976): 237–249.

7.4. H. Carrillo, The method of averaging and stability under persistent disturbances with applications to phase-locking, Dissertation, University of Utah, 1983.

7.5. C. Moler, On the calculation of exponentials of matrices, *SIAM Review*, 1980.

7.6. F.C. Hoppensteadt, W.L. Miranker, Multi-time methods for systems of difference equations. *Studies Appl. Math.*, 56(1977): 273–289.

7.7. W. Feller, *An Introduction to Probability Theory and its Applications*, Wiley, New York, 1968.

7.8. R. Novick, F.C. Hoppensteadt, On plasmid incompatibility, *Plasmid* 1(1978): 421–434.

7.9. J.K. Hale, *Oscillations in Nonlinear Systems*, McGraw-Hill, New York, 1963.

7.10. N.N. Minorsky, *Nonlinear Oscillations*, Van Nostrand, Princeton, 1962.

7.11. N. Krylov, N.N. Bogoliuboff, *Introduction to Nonlinear Mechanics*, Annals of Mathematics Studies, No. 11, Princeton University Press, Princeton, 1947.

7.12. N.N. Bogoliuboff, Y.A. Mitropolski, Asymptotic Methods in the Theory of Nonlinear Oscillations, Gordon-Breach, New York, 1961.

7.13. J. Moser, *Stable and Random Motions in Dynamical Systems: With Special Emphasis on Celestial Mechanics*, Princeton University Press, Princeton, 1973.

7.14. J.D. Cole, *Perturbation Methods in Applied Mathematics*, Blaisdale, Waltham, MA, 1968.

7.15. J. Kevorkian, J.D. Cole, *Perturbation Methods in Applied Mathematics*, Springer-Verlag, New York, 1981.

7.16. S. Wiggins, *Introduction to Applied Nonlinear Dynamical Systems and Chaos*, Springer-Verlag, New York, 1990.

7.17. P.R. Halmos, *Measure Theory*, Van Nostrand, Princeton, 1950.

7.18. J. Moser, On the theory of quasi-periodic motions, *SIAM Rev.* 8(1966): 145–171.

7.19. H. Goldstein, *Classical Mechanics*, Addison-Wesley, Reading, Mass., 1950.

7.20. J. Grasman, E.J.M. Velig, G. Willems, Relaxation oscillations governed by a van der Pol equation, *SIAM J. Appl. Math.* 31(1976): 667–676.

7.21. M.A. Cartwright, J.E. Littlewood, *Ann. Math.* 54(1951): 1–37.

7.22. J.E. Flaherty, F.C. Hoppensteadt, Frequency entrainment of a forced van der Pol oscillator, *Studies Appl. Math.* 58(1978): 5–15.

7.23. J.L. Lions, A. Bensoussan, G. Papanicolaou, *Asymptotic analysis for periodic structures*, North-Holland, New York, 1978.

7.24. M.E. Munroe, *Introduction to Measure and Integration*, Addison-Wesley, Cambridge, Mass., 1953.

7.25. F.C. Hoppensteadt, W.L. Miranker, An extrapolation method for the numerical solution of singular perturbation problems, *SIAM J. Sci. Stat. Comp.* 4(1983): 612–625.

7.26. W.L. Miranker, *Numerical Methods for Stiff Equations and Singular Perturbation Problems*, D. Reidel, Holland, 1981.

7.27. F.C. Hoppensteadt, A. Schiaffino, Stable oscillations of weakly nonlinear Volterra integro-differential equations, *J. reine u. angewandte Math.* 353 (1984): 1–13.

7.28. D.S. Cohen, F.C. Hoppensteadt, R.M. Miura, Slowly modulated oscillations in nonlinear diffusion processes, *SIAM J. Appl. Math.* 33(1977): 217–229.

7.29. S.C. Persek, F.C. Hoppensteadt, Iterated averaging methods for systems of ordinary differential equations with a small parameter, *Comm. Pure Appl. Math.* 31(1978): 133–156.

8.1. F.C. Hoppensteadt, Singular perturbations on the infinite interval, *Trans. AMS*, 123(1966): 521–535.

8.2. F.C. Hoppensteadt, Properties of solutions of ordinary differential equations with small parameters, *Comm. Pure Appl. Math.* 24(1971): 807–840.

8.3. R.E. O'Malley, *Introduction to Singular Perturbations*, Academic Press, New York, 1974.

8.4. M. van Dyke, *Perturbation Methods in Fluid Mechanics*, Parabolic Press, Palo Alto, C., 1975.

8.5. J.D. Cole, *Perturbation Methods in Applied Mathematics*, Blaisdale, Waltham, Mass., 1968.

8.6. J. Kevorkian, J.D. Cole, *Perturbation Methods in Applied Mathematics*, Springer-Verlag, New York, 1981.

8.7. J.J. Stoker, *Nonlinear Vibrations*, Wiley, New York, 1950.

8.8. K.O. Friedrichs, Asymptotic phenomena in mathematical physics, *Bull. AMS*, (1955): 485–504.

8.9. A.B. Vasil'eva, Asymptotic formulae for the solution of a system of ordinary differential equations containing parameters of different orders of smallness multiplying the derivatives, *Dokl. Akad. Nauk SSSR*, 128(1959): 1110–1113.

8.10. W. Wasow, *Asymptotic Expansions for Ordinary Differential Equations*, Interscience, New York, 1965.

8.11. L.E. Frankel, On the method of matched asymptotic expansions, *Proc. Camb. Phil. Soc.* 65(1969): 209–284.

8.12. J.L. Massera, Contributions to stability theory, *Ann. Math.* 64(1956): 182–206; 68(1958): 202.

8.13. A.B. Vasil'eva, V.F. Butuzov, *Asymptotic Expansions of Solutions of Singularly Perturbed Equations*, Nauka, Moscow, 1973 (in Russian).

8.14. G.E. Briggs, J.B.S. Haldane, A note on the kinetics of enzyme action. Biochem J. 19(1925): 338–339.

8.15. M.A. Cartwright, J.E. Littlewood, *Ann. Math.* 54(1951): 1–37.

8.16. N. Levinson, A second order differential equation with singular solutions, *Ann. Math.* 50(19): 127–152.

8.17. L.S. Pontryagin, Asymptotic behavior of the solutions of systems of differential equations with a small parameter in the higher derivatives, *AMS Transl. Ser. 2*, 18(1961): 295–320.

8.18. E.F. Mishchenko, Asymptotic calculation of periodic solutions of systems of differential equations containing small parameters in the derivatives, *AMS Transl. Ser. 2*, 18(1961): 199–230.

8.19. J. Grasman, E.J.M. Velig, G. Willems, Relaxation oscillations governed by a van der Pol equation, *SIAM J. Appl. Math.* 31(1976): 667–676.

8.20. J.E. Flaherty, F.C. Hoppensteadt, Frequency entrainment of a forced van der Pol oscillator, *Studies Appl. Math.* 58(1978): 5–15.

8.21. M. Levi, F.C. Hoppensteadt, W.L. Miranker, Dynamics of the Josephson junction, *Quart. Appl. Math.* (July 1978): 167–198.

8.22. P. Kokotovic, H. Khalil, *Singular Perturbation Methods in Control: Analysis and Design*, Academic, London, 1986.

8.23. N. Gordon, F.C. Hoppensteadt, Nonlinear stability analysis of static states which arise through bifurcation, *Comm. Pure Appl. Math.* 28(1975): 355–373.

8.24. S. Chandrasekhar, *Hydrodynamic and Hydromagnetic Stability*, Oxford University Press, 1961.

8.25. P. Alfeld, F.C. Hoppensteadt, *Explosion Mode Analysis of* H_2—O_2 *Combustion.* Chemical Physics Series, Springer-Verlag, New York, 1980.

8.26. N.N. Semenov, *Chemical Kinetics and Chain Reactions*, Clarendon, Oxford, 1935.

8.27. A.C. Hindmarsh, Gear's ordinary differential equation solver, UCID-30001 (rev. 3) Lawrence Livermore, Lab, Livermore, CA, Dec. 1974.

8.28. W.L. Miranker, *Numerical Methods for Stiff Equations and Singular Perturbation Problems*, D. Reidel, Holland, 1981.

8.29. F.C. Hoppensteadt, W.L. Miranker, An extrapolation method for the numerical solution of singular perturbation problems, *SIAM J. Sci. Stat. Comp.* 4(1983):

8.30. R.Z. Khas'minskii, *Stochastic Stability of Differential Equations*, Sijthoff & Noordhoff, Rockville, Md., 1980.

8.31. M.I. Freidlin, A.D. Ventsel, *Random Perturbations of Dynamical Systems*, Springer-Verlag, New York, 1984.

8.32. F.C. Hoppensteadt, *Mathematical Theories of Populations*, SIAM Publications., 1975.

8.33. J.M. Greenberg, F.C. Hoppensteadt, Asymptotic behavior of solutions to a population equation, *SIAM J. Appl. Math.* 17(1975): 662–674.

8.34. F.C. Hoppensteadt, On quasi-linear parabolic equations with a small parameter, *Comm. Pure Appl. Math.* 24(1971): 17–38.

8.35. E.N. Lorenz, Deterministic nonperiodic flow, *J. Atoms. Sci.* 20(1963): 130–141.

Additional recommended reading:

V.I. Arnol'd, *Mathematical Methods of Classical Mechanics*, Springer-Verlag, New York, 1978.

K.G. Beauchamp, *Walsh Functions and their Applications*, Academic Press, New York, 1975.

P.H. Carter, An improvement of the Poincare-Birkhoff fixed point theorem, *Trans. AMS* 269(1982): 285–299.

J. Guckenheimer, P. Holmes, *Nonlinear Oscillations, Dynamical Systems and Bifurcations of Vector Fields*, Applied Mathematical Sciences, Vol. 42, Springer-Verlag, New York, 1983.

P.R. Halmos, *Lectures on Ergodic Theory*, Chelsea Publishing Co., New York, 1956.

J.M. Hammersley, D.C. Haudscombe, Monte Carlo Methods, Methuen, London, 1964.

E.L. Ince, *Ordinary Differential Equations*, Dover, New York, 1956.

E. Jahnke, F. Emde, *Tables of Functions*, Dover, New York, 1945.

H. Jacobowitz, Corrigendum, The existence of the second fixed point, *J. Differential Eqns.* 25(1977): 148–149.

W. Leveque, *Elementary Theory of Numbers*, Addison-Wesley, Reading Mass., 1962.

W.C. Lindsey, *Synchronization Systems in Communication and Control*, Prentice-Hall, Englewood Cliffs, N.J., 1972.

Index

Applied Mathematical Sciences

(continued from page ii.)